Fred Red Cell
and
Friends

Mike Pearce

DEDICATION

This book is dedicated to the hundreds of students that had to suffer my enthusiasm for human biology and went on to health professions. Also to my family and grandchildren for suffering the 'did you knows?' of which there are many in this book. Thanks go to my wife for proof reading. Finally to those animals that never achieved the ability to develop parental care and social organisation left their young to cope on their own from the start.

CONTENTS

CONTENTS CONTINUED

Preface

There are hundreds of books on human biology which state facts in many different ways. More recently glossy photographs or coloured diagrams have appeared to help students relate pictures to the text. All these, with the internet, tend to form good fodder for the student to churn out essays.

When a body system is explained, students are often bombarded all at once with hundreds of new names. The next body system studied tends to have a completely new set of names and so it goes on. This can, for some students, produce a huge barrier or turn off to learning about the basics behind the physiology and function of the human body. However, the odd or unusual piece of information seems to be remembered more than the more mundane facts. This piece of information can often form the key or link to access and remember the rest.

This book as an introduction has gone several steps further, some may say way out, or too extreme in order to help students easily visualise and relate the wider picture of the body systems. It gives organs and other body parts people's first names and also puts in snippets of related facts for other animals as comparisons.

Added at the end of some chapters are examples of some common diseases and disorders and how these may disturb the balance of the body or have even more devastating effects on health. Also in some chapters the effects of ageing on the body is examined as you all get old in the end.

This book forms a great basis for GCSE and A level courses, Diploma, NVQ/Apprenticeships and vocational courses as well as Access courses for Human Biology for Health Studies.

It also provides the lay person with information on how their body works and what can go wrong as well as showing the amazing structure and relationships which exist in you as a multicellular organism existing in the foreign environment of this world.

It also has a wealth of information on how, when, where, biggest and smallest for lecturers and teachers which can be used to make lectures a little more interesting and enriching and often includes comparisons with the animal world who are often more advanced in some respects. With all publications like this information always changes and new facts and figures emerge all the time. Some theories remain theories or are changed. New records and feats are exceeded. The author apologises for any errors or unbelievable items that may have crept into this publication, but is sure that its usefulness overrides any mistakes that exist and hopes you enjoy reading this book.

Let's start with Fred

I'm Fred the Red Blood Cell who was born in a Billy Bone

I never stop moving and don't have a home.

I'm destined to travel from your head to your toe

To tell you my tales wherever I go.

Each tissue or cell has been given a name

Please don't get offended if yours is the same! (e.g. Tony Testes, Rita the Ureter and Rosie Rectum.

FRED RED CELL AND FRIENDS

It's a Banjo sandwich some will say

Stuffed full of facts that won't go away

So give it a go, come for a ride

Jump into a cocktail of facts and let Fred be your guide.

1 WHO AM I?

Most biology books start with 'what is a living organism'?
This is defined usually as something which moves,
breathes, grows, reproduces, excretes, responds and feeds.
Really none of these things would happen in most
multicellular organisms unless there is a pump to transport
food and waste materials around so they don't clog up the
system. Where there is water there is life and life depends
on fresh water. (Your body is eighty percent water.) Also
we need a pump which is suddenly kick started in Happy
heart Mollie muscle cells on day twenty two inside your
mother and starts my journey.
Hi I'm Fred the red blood cell one of the many important
items this little Happy heart pumps around, and one of the
fifty thousand million cells in the body, and some of the
ninety trillion atoms that make up the body. My colour is
really brown to dark purple but blush bright scarlet when
filled with oxygen which is seen when you bleed.
I've been around a long time, over five hundred and
seventy million years since life appeared. Ninety-nine per
cent of all animal species became extinct before the human
species came on this earth. Ninety nine percent of all
creatures that existed have not been found in fossil records
but likely contained some of my brothers. Unfortunately,
many of these were soft bodied so had no skeleton so you
don't know about them from fossil records. Flatworms
and corals don't have blood so you would never find me
there. I was around long before you existed when your
ancestors were a 'subordinate' species at the time of the
dinosaurs and you only came out at night to scavenge. You
have only been around two million years. The dinosaurs
were around for a hundred million years .With ice ages,

water was locked in ice and caves provided protection for all animals and us cells deep inside you.

I'm a born traveller, whizzing round your body one thousand four hundred and forty times per day (i.e. once a minute!) I am in a total blood volume of 4 to 6 litres. Women have 4.5 litres while men have 5.6 litres of blood. Eight percent of your body is blood, fifty percent of which is blood plasma. I have the potential to travel hundred thousand kilometres which is the estimated length of all the blood vessels in the body and two and a half times around the earth. I can take twenty seconds to travel around the whole length of the body but that's pushing it, and I need to be on a fast highway. Size-wise about a thousand of me and my brothers fit onto a full stop on this page. Of the two hundred and fifty million blood cells in a drop of blood five million of us Fred red cells exist (compared to ten thousand Winston white cells and two hundred and fifty thousand Pluto platelets). Over twenty thousand billion of us may be in your body at any one time compared to a total of seventy five trillion cells in your body. I can change shape, as blood is not very thick, I am a good squeezer through teeny blood vessels. I can also bunch together with my mates and stick together as a clot. We are at maximum numbers at around twelve noon every day waiting to take away your broken down food and pick up oxygen in order to convert this food to energy.

I was born in Billy bone marrow, the jelly enjoyed by dogs and cavemen alike. We can reach our largest size of 0.2 millimetres. Three million of us are made per second; two hundred million of us are destroyed per minute, wow! (More if we injure ourselves or oxygen level falls in your body). There are more of us in males than females. If you are living a thousand metres above sea level there are extra millions of me as more are needed as oxygen is in short supply.

I have a thin Slinky skin like membrane (made of fat and phosphate with currant like proteins embedded in it). Liquids, gases, salts and a lot of other chemicals can move in and out of my membrane and also through holes in my membrane protein currants. The proteins also give me a label (identity) useful if other killer cells are after me.

Diana DNA is in all cell centres except mine! Originally when born I had Diana DNA (Jenny genes stuff) in my nucleus like Fred red cells of frogs and birds which still keep a nucleus. Some animals have more Diana DNA than I originally had e.g. the one celled jelly like Amoeba. Now I don't have Diana DNA so I cannot divide like the other cells when you are asleep! Diana DNA would also have allowed me to transfer genetic material, pretty useful for coding for what you look like etc. and what your children will look like. Sadly alas I lost my nucleus and became filled with iron in order to do another job. There is enough iron in your body to make a three inch nail and set off a metal detector. I'm also packed with a protein called globin (not Goblin) which is attached to this iron (haem) to give the famous haemoglobin found in Fred red cells. Rust is red so the rusty iron haem in my cells makes me red - often seen when you cut yourself.

With the rise of the plant kingdom on the earth and in the sea the oxygen in the air increased and animals and plants could grow bigger. Everything with iron in (my haemoglobin) which included me started rusting. However, my contents are recycled when I die to help produce others like me which is something I suppose. Men, as they have more cells, have more haemoglobin in their blood than women. So that's why they talk about full blooded males? In blood vessels I swim in a bath of straw coloured fluid called plasma, (two to three litres in all.) which carries everything from food (fats, proteins, and sugars), water (seven percent of the body's water), salts, waste products, not to mention hormones, drugs and alcohol.

Swimming with me are my closest relatives 'Winston white blood cells'. These are the body's soldiers, whose job is to eat and pop Bronwyn bad bacteria, and fight Veronica viruses, produce antibodies and develop into killer and memory killer cells? Winston white cells will even destroy me in Lily liver and Spiro spleen when my life is over (Real terminators). Winston white cells are most abundant at 11pm each day (Bronwyn bacteria come round for elevenses, we're ready to get you).

I travel round the body in different sized tubes (blood vessels) which end to end would travel two and a half times round the earth - quite a lot really? I zoom through Ashley arteries under high-pressure but only trickle in Vernon's veins. Travel is even slower in Christine's capillaries (fine tubes) where I queue in single file, bit like waiting for a pop concert or a football match – only no overtaking here! Even squeezing won't get me through.

In evolution, for those that believe in it, (a lot still don't) we were all once single cells. A few of us got together to form a gang (cell balls or tissues). Cells could capture food and take it in from the outside and also excrete waste. When the gang got too big they needed tubes to get food and gas around them and remove waste and most of all a muscly pump your Happy heart, to move the lot through the tubes
On my travels in the tubes in your body I am able to meet even more cell relatives. Two hundred different cells exist. The biggest cell that exists is the Elsie ostrich egg cell. In humans, one of your cells has been stretched long and thin, others are small with many connections as found in your Brian brain - this is the 'Nellie's nerve big family, made up of twenty five billion Nellie nerves!

Mollie muscle cells in the arms, legs and Gillie gut, have also been stretched poor things, but into long bundles of sliding fibres. In Happy heart these cells are pulled into a fishnet structure which helps different parts contract at the same time. My friends and I are often in here carrying sugar oxygen to these fibres.

Another of my relatives is Bob Billy bone builder. They do this by secreting calcium and phosphorus (hard stuff). Also another relative is Danny demolition cells which dissolve this hard stuff in Billy bones away to supply calcium to other regions of the body when needed, especially if you are pregnant and the baby needs to make Billy bone. Too much removed and you are in trouble.

Some of my other cell friends shape wise are themes on a square shape. Ellie epithelium is one of them (Your flaky outside Slinky skin). These cells are as flat as pancakes or like thin wafer pavement slabs. You will need a duster as most of your Slinky skin surface is lost every two weeks (Forty pounds in a lifetime). This kind of Ellie epithelium also lines the inside of Gillie gut and is changed every five days.

Other epithelial cells are shaped like sugar cubes or are rectangular with or without finger like extensions called Victor villi. Where these tentacle like little fingers exist this may be on the outside of cells found in the inside of your Gillie gut which increase the area for absorption of liquid food as well as moving food along. This way Mr Burger, your Victor villi await you. These Ellie epithelial cell may also be hairy on top (Cecilia's cilia), ready to waft up the gunge in your Lucky lungs to cough up or swallow if you do not kill Harry hairs first by smoking. A similar set of Harry hairs is found in the female's oviducts which are tubes which carry Elsie egg down to Wendy womb. As a blood cell I also travel to your organs that produce sex cells, Olive ovaries with Elsie eggs in the female and Tony testes which produce Speedo sperm in the male.

Your animal cells, unlike plant cells, do not have the green pigment chlorophyll which captures energy from sunlight to convert sugars to energy for growth etc. This is a shame because if you had a green head you could absorb sunlight when sunbathing or walking in the street and make sugar for energy and could eat less food and save the fossil fuels - You never know, a bit of genetic engineering putting back the photo cells into you could change everything. However, not much use on a cloudy day.

Other cells in the body include packing cells (areolar tissue). These stop your internal organs swinging around like bells which could be damaged when you wake or exercise. Another example of packing cells, well known to those on a diet, is Freda fat cell (adipose tissues). She is so stuffed with her yellow fat that her nucleus, normally in the centre of the cell is shoved to one side like a diamond on a ring. You have over forty five billion Freda fat cells in your body and can have more. Not all your fat is in Freda's cells (your brain is like a lump of fat –Nellie nerve fat). Your bodies have enough fat to make seven bars of soap (quite a lather).

Lastly you may find busy Susie secretory cells which produce exotic fluids when and where required. Examples of secretions include messenger hormones or reactive enzymes, in little packets, which when released travel in my plasma to the place where they are to perform (like little actors in a play waiting to go on stage). Other Susie secretory cells include those in Gladys sweat glands (mostly on the hand).Gladys Sweat glands are responsible for sweaty brows and smelly arm pits, Gladys tear glands for crying (not salty crocodile ones) also help to kill remove dust and micro-organisms in the eye. In space there are no tears just a little pool at the base of the eye.

There are over four thousand Ewan ear wax glands.
Hearing aids can increase wax secretions which
unfortunately block the Ewan ear hole. Pardon me Sir I
didn't quite get that, I still need the subtitles on TV.

My cell friends in blood and the fluid plasma supply all
organs in the body including Happy heart, Lucky lungs,
Lily Liver, Katy kidneys, Brian brain and reproductive
organs, with everything to keep them going. We need to
make sure there is around thirty percent of the body's
water around each body cell and in the lymph and sixty
percent water actually inside the cells. All cells, except
myself, have little organelles (teeny organs) inside them.
Some of these help with respiration like mitochondria
(previously thought to be a Bronwyn bacteria) to produce
energy.
Some organelles make proteins or fats and other useful
stuff. These materials are put into packets to transport
across the cell membrane to travel in my blood.
Transport can be simple movement from high to low
concentrations i.e. diffusion. (High to low water
movement through a partially permeable membrane is
called osmosis.) Other methods to get fluid in and out of
membranes include exocytosis (fluid in) and endocytosis
(fluid out) where little packets enter or leave by the
membrane. You can also use energy to pump things across
membranes which is called active transport. This is useful
when you want to get back to normal.

Three hundred million cells die every minute in the body
but some cells, like Billy bone, can live for thirty years
before being replaced. By the time you have read this
sentence over fifty thousand cells in your body have died
and been replaced. I can travel over two hundred and fifty
times round your body in my lifetime and I can live as long
as three to four months if I'm lucky. Some cells live longer
depending on their environment. It is believed that animals

living in water live longer as they have less stress from gravity and they are cold blooded animals so cell processes are slower.
So now you know where I was born, met some of my relatives which form tissues and organs and know how I differ from other cells.

Let's go for a 'bloody' ride

2 HAPPY HEART PROVIDES A TICKET FOR FRED RED CELLS ROUGH TUBULAR RIDE AROUND THE BODY

The circulatory system consists of Happy heart, blood vessels and Lesley lymphatic system. Happy heart was thought in the Middle Ages to be the centre of intelligence. Even today some think that the nervous system of Happy heart is linked somehow in this way. Happy heart lies close to Lucky lungs in the middle of Charlie chest but slightly to the left. If you are a shrimp then it's in your head. Don't let your Happy heart get to your head?

Some people are born with two Happy hearts like Dr Who but this is very rare. After three weeks the human embryo cannot depend on diffusion any more to obtain nutrients so early Christine capillaries are formed. At this time two tubes join together to form Happy heart but cannot expand length wise so expand sideways giving Happy heart shape. Also more importantly, around the same time a miracle of life occurs. Happy heart's tubes are present, and then something chemical/electrical in a small part of the Mollie muscle suddenly triggers Happy's first Happy heartbeat. Initially this starts as little waves of contraction (peristalsis) but then becomes as single waves. The first beat is often heard at fifteen to twenty two weeks. True life has started.

Circulation is a double system. Blood from Happy heart goes to Lucky lungs and then back to Happy heart. This is useful for fast moving creatures like mammals where a lot of oxygen is needed to burn food to make energy to run super-fast. This makes sure you get the best food, catch a

beautiful mate or run away fast from a beastly predator or even a wasp that wants to sting you.
If you had long necks you would need to pump blood upwards. Giraffes' have happy hearts two feet long! To achieve this that's quite a Happy heart? Good for biting apples on trees (the neck that is.) Some animals e.g. octopus have three Happy hearts. They have lots of legs so may need this. Happy heart can be on the right side in some people and not on the left as normal.

Happy can renew itself, but this is slower than your Slinky skin. Happy in humans is fist sized, males tend to have bigger Happy hearts than females (structurally speaking), but women's Happy hearts beat faster than men's Happy hearts. Happy inside has four chambers (room like). It sits between Lucky lungs, pumping, pumping, and pumping, never resting. It beats at rest seventy two beats per minute hundred thousand times per day and beats over two billion times by the age of seventy! It does stop when you sneeze but not for long. The smaller an animal the faster the lifestyle and therefore Happy heartbeat. The shrew's Happy heart beat is one thousand two hundred per minute. Happy hearts power in humans is one hundred times less than a car engine but still is very effective. Happy pumps out seventy millilitres of blood at each beat, ten thousand litres a day. It is said this could in seventy years fill three super tankers, or a hundred Olympic swimming pools. In animals in water Happy heart rate slows down as does the breakdown of food for energy production.

Ashley's arteries always leave Happy heart, and Vernon's veins enter him. As a red blood cell I am red in colour but in crustaceans and some spiders my equivalents are blue due to copper, not iron inside these cells.
With my fellow red blood cells we then go on a rough journey. We flow from the body into the largest of Vernon

veins into one of the top rooms (right atrium) of Happy heart and are forced into one of the bottom two chambers (right ventricles) Here we pass through the valves from which we are forced round Lucky lungs (via Ashley's pulmonary artery). We then flow back via the pulmonary vein into the other top chamber (left atrium). Blood then enters the other bottom chamber. I leave the ventricle at very high pressure through the Ashley's biggest vessel, the aorta, (three centimetres wide, as big as your thumb) to go round the head and body. In a baby, there is a hole between the two top chambers so venous blood is mixed blood and goes to the lower chambers. The hole seals up later. Ashley arteries are deeper in the body than Vernon veins so as to protect them. Sixty five percent of the blood at any one time is in Vernon veins.

Happy heart, like all pumps, needs valves (like doors to rooms) between his chambers, to stop backflow (like some beach inflatables.) The lub dub sound heard with a stethoscope is the closing and opening of these valves when blood pressure is measured. The inflated cuff allows blood in the artery to flow past for only a part of each Happy heartbeat; this causes sounds in the artery. The cuff is inflated until it stops blood flow when the first sound is heard, this is systolic blood pressure. When sound disappears the cuff gradually deflates - this is diastolic pressure which is the pressure when Happy heart relaxes again.

Your blood pressure varies with age but in young adults is 100-120 millimetres mercury (Hg). The systolic blood pressure is the maximum coming from the aorta (usually 100 added to your age). Diastolic blood pressure is when Happy heart is resting. This normally 70-80 millimetres Hg (mercury) or 80-90 in 40-60 year olds). When represented the systolic blood pressure value is given then the diastolic. The lower your blood pressure the better. Higher than 140 over 90 means you could be at more risk of Happy heart problems. Blood pressure is supposed to be highest at 6.30 pm, with its sharpest rise at 6.45 am. Blood is at its thickest at 6-8am though this varies for some people. Highest blood pressure is seen in a giraffe to get it up the neck. In some countries no operations are done when the full moon occurs as blood pressure increases with aggressive behaviour which is believed to occur at that time. Blood volume is lower in winter than in summer and blood flow in women is higher at night than in the morning.

Blood pressure is often so high that it expands these muscular Ashley arteries. Happy heart is a big squirter. He can squirt blood up to nine metres high (red ceilings for artery throat cutters). Rushing through Happy heart for me as a red blood cell is like being on a helter skelter fairground ride sending me and millions of others suddenly out through the huge aorta tube to the head or around the body. The wedding ring was placed on the fourth finger of the left hand as it was thought there was an artery that ran straight to Happy heart.

Your pulse shows this high pressure as a wave running through the blood and can be felt best at your wrist or over the carotid artery on your neck. Happy heart beats for a man at rest/sitting is seventy to eighty per minute, for a woman seventy eight to eighty two and children eighty five per minute. Often Happy's heart beat follows the body's clock (circadian rhythm). A hedgehog's Happy heart beats three hundred times per minute probably because it moves fast and travels a lot, several gardens a night.

Many factors can affect your pulse readings, the very act of having your pulse taken can be a factor as well as the effect of seeing certain colours. Reds, yellow and orange are supposed to stimulate your pulse rate. Possibly this is why when women wear red they are noticed more or want to be noticed by men. An increase in temperature may also means Happy heart has to work harder as Vernon veins dilate. Sudden violent exercise can raise the beats to as high as two hundred per minute. However with training a sprinters Happy heart rate can be lowered to sixty beats per minute, a distance runner forty five beats per minute and a marathon runner forty beats per minute.

After exercise in a non-trained person the rate takes a few minutes to return to normal, but with a lot of sudden exercise this can take several hours. The blood has to get rid of the excess lactic acid produced by Mollie muscles. Happy heart itself, as well as using glucose and fatty acids for energy, uses lactate, a milk sugar.

As well as passing through the inside of Happy heart us Fred red cells actually pass over and through Mollie muscle in Happy heart's walls (the old net) as she also needs oxygen and glucose and fatty acids for energy so as to squeeze the blood from chamber to chamber and out. To make sure Happy heart does not pump too fast or slow she has a pacemaker (a timer or beat manager). This is a bundle of Nellie nerves like a small watch battery that generates electrical signals. These little shocks leak sodium at a pre-set rate (a little shock goes a long way). Shocking but true, this starts and keeps the top chambers of Happy heart (atria) contracting. An electrical signal (shock) from this pacemaker then passes the signal (shock) to another bundle of Nellie nerves to the bottom chambers to start them contracting. So shock top first (atria) then shock bottom next (ventricles) brief rest then repeat. Many other animals e.g. sea hares, have a similar electrically generating shocking cell. Basically these electric cells are leakier to positively charged atoms (ions) i.e. more sodium in more potassium out, makes your Mollie muscles jump about.

On my way round Happy heart as Fred the red blood cell I may meet different kinds of body monitoring detectors-receptors which detect change. Baroreceptors monitor pressure). Stretch receptors monitor the amount of Mollie muscle extension. The amount of acidity etc. in the blood is checked by chemoreceptors in the chambers and large blood vessels of Happy heart. Chemoreceptors also check on how much carbon dioxide dissolved in the blood. If there is too much the brain gets Happy heart to beat faster so as to get rid of some carbon dioxide and pick up more oxygen in Lucky lungs.

In your body you can have over a hundred thousand kilometres of blood vessels (6.5 square centimetres of Slinky skin has over six metres of blood vessels). Inside Ewan ear there is no pulse as the noise would affect your hearing. However blood does pass through the Ewan ear drum and sometimes causes a kind of pulse.
Oxygenated blood gives us red cell their scarlet role. This passes around the body in Ashley's arteries (2- 30 centimetres per second depending on closeness to Happy heart) under pressure and goes up to the head and Brian brain. It seeps through small Christine capillaries before going into the Vernon veins and back to Happy heart. We then look a bit dull, a rather dark rusty colour. Lack or exercise and a sedentary life causes less blood oxygen and gives a more blue appearance to the blood. This gave rise to the term 'Blue blooded' where some of the royalty in the past were not very active (Couch potatoes).

Once we Fred red cells trickle down to your feet, gravity grabs us down and we have a problem getting back up your legs in Vernon veins. Please get us back up, all is forgiven. If you put us on the moon we would have no problems there except getting back down. (No gravity). When astronauts arrive back on earth they also faint as their circulation has not adapted yet to gravity as before. On earth four hundred to six hundred millilitres of blood pools in the lower extremities, legs and feet. This means there is higher blood pressure in the feet than in the head.

Mollie leg muscles however, come to the rescue giving
Vernon veins a squeeze to get us back up. However, if the
body we are in stands still for a long time, as with soldiers
on parade, Mollie muscles cannot move and squeeze
Vernon veins so we cannot get up the legs very fast, so less
blood goes to Brian brain. You will collapse onto the floor
(fainting), so please lie down so we can flow again. A tight
collar can also have the same effect, restricting blood
supply and therefore oxygen to Brian brain. Loss of
consciousness can be followed by loss of kidney function,
blood flow and filtration.

Blood flow in Vernon veins in the body therefore
depends a lot on squeezing by Mollie muscles Bloodletting
was very common in the past. Losing us blood cells may
have woken up and triggered the body's defence system.
The shop where bloodletting commonly occurred was
represented by the barber's pole (red for blood and white
stripes for bandages). The ball at the top end represents
the basin for collecting the blood.
Even monks removed blood as they thought it would
remove worldly thoughts. People also used leeches and still
use them today to reduce bruising especially on the face
and around the eyes. Leeches also, if put on the tip of an
amputated finger which has been reattached help to bring
the blood back up to help revive the circulation. In the
Second World War blood transfusions used coconut milk
instead of plasma. I am sure this gave a shock to the
system but it worked.

In Vernon's veins, unlike Ashley's arteries, there are valves
which slow us blood cells down and help us move
forwards or back against gravity by preventing back flow.
The deep Vernon veins are massaged by the calf Mollie
muscles. Valves are absent however in Vernon veins in the
brain; blood just drains back to Happy's heart by gravity.
When Ashley's arteries are cut in the body the Mollie

muscles try to close up to stop the bleeding. Giraffes have valves in Ashley's arteries in the neck so blood supply to the neck does not flow back, so reaches the head.

In Christine's capillaries we squeeze by each other or may move in single file so that we and your plasma can swop goodies (including minerals and vitamins) and off load baddies (waste) and gases. Christine's capillaries make sure they keep the good stuff inside them e.g. big proteins as they cannot pass through. There are around ten billion Christine capillaries in total (around sixty Kilometres) in the body the largest having a diameter of 0.2 millimetres and many being 0.01 millimetres. In some parts of the body there are many of Christine's capillaries close together forming a fine network (a Christine's capillary bed). These are in areas where there is a lot of exchange of materials.

Hitting Slinky skin makes it go white as blood is removed. The body thinks a lot of blood loss is coming and causes blood vessels to shrink so less blood is sent to this area. Then the area becomes bright red as a greater response occurs to get blood back to that area to fix any damage. Blushing is the expansion of Christine's capillaries in the face; the same is seen after drinking alcohol. Sometimes these remain giving thin thread lines on the face. We also supply very small or thin areas in the body such as in Ellie eye, the middle layer, the ciliary body and the iris. We do not supply the outer transparent layer of Ellie eye which is the cornea as this would blur your vision. To keep this thin membrane alive food and minerals diffuse through, in and out of the membrane.

Too much oxygen in areas of the body can reduce Christine's capillary growth but if less oxygen is supplied after this it has the effect of overcompensating by producing more Christine capillaries. An increase in Christine's capillaries is associated with tissue growth associated with cancers, wound healing, and Una uterus expansion.

Babies in Wendy womb have two Ashley's arteries and only one Vernon vein in their umbilical cords which makes sure a good flow of nutrients arrive in Ashley arteries from the mother through Penelope placenta.

As a red blood cell I hate being in an acid or alkaline bath, but I'm saved by Amino acids. These in your blood proteins help to regulate the blood's acidity and alkalinity i.e. they buffer the changes by changing their chemistry. If the pH reaches 6.8 - 8 that's alkaline, you're dead as it changes charges on cells especially Nellie's nerves. This causes you to spasm and Mollie's muscles for breathing don't work. If your blood is too acid this affects Brian brain, making you more and more depressed as well as killing off enzyme activity so reactions in the body cease. Salt also can affect the level of electrolytes in blood.

Blood flow is important in temperature control. This depends on a counter current system i.e. I am whizzing one way in a vein while my friends are whizzing close by in the opposite direction and can give or take any excess heat. Heat can be lost from my plasma by convection and conduction from the plasma. The rate of blood flow is also linked to temperature and determines how much heat is lost. These relationships evolved in the past when we blood cells flowed around in reptiles including the dinosaurs. That's why some had big fin like floppy sails on their backs which some lizards still have today. This is also linked to elephant Ewan ears and birds standing on a cool surface where the heat exchange mechanism comes in. This involves shunting blood to different areas to help lose or gain heat. All humans have Noddy nose which can be involved in heat exchange. Slinky skin and excretory products wee and poo also help to lose heat. Sweating keeps you cool and in normal conditions you may lose a pint a day. Some animals don't sweat but remove heat through panting. Reindeer eat moss which has antifreeze and is supposed to help keep them warm.

Another system closely linked to the circulatory system is Lesley Lymphatic system. This has as many vessels in your body just like Christine capillaries. Lesley drains and filters using sieve like structures in nodes and cleans your blood of all debris and muck as well as keeping tissue fluid flowing and stopping flooding between cells. Lesley lymphatic also produces white blood cells and antibodies which fight off external invaders. Laughing is said to boost up your antibodies and help improve your immune system so a few jokes can help you live longer. Perhaps you should not say you are killing me with that joke, it's so funny and makes me laugh.

One to two litres of lymph is circulated through the vessels and tissues in your body. The largest Lesley lymphatic tissue is Spiro spleen which is hidden under your lower ribs below your Dickie diaphragm and left of Stella stomach. It is dark purple and looks like a spongy baseball glove and is a major organ for fighting infection and killing Bronwyn bacteria. It also can produce antibodies against viruses. It produces some blood cells in the foetus before birth and removes old Fred red cells. Other examples of Lesley's lymphatic tissue include Terrence tonsils and Aden adenoids and Tanya thymus, are important for defence. Lose them and you are less defenceless. Enlargement of Lesley's lymphatic tissue can be caused by infections hence the term swollen Gladys glands. In some cases these can reach the size of an orange!
 One bad thing about lymph nodes is that they provide a means of transport for cancer cells to other parts of the body and also a place where the HIV veronica virus can multiply, spread and kill your defence cells and lower your immune system.

21

Circulatory system problems

If your blood pressure is over 160/90 you have hypertension. If higher still there may be other risk factors responsible e.g. smoking. Salt affects blood pressure. Without salt you would die. Too much salt in your body means more enters Mollie muscle cells blood vessels allowing more calcium in so that they contract and narrow the walls. You may need a check-up.

Hypertension can indicate a blockage slowing the blood flow and more fluid can collect in Lucky lungs. Complications can occur but this is often when some of the damage has already been done. Hypertension and also diabetes is more common in Afro Caribbean people as it is believed they may be less tolerant to high salt intake. They have lower levels of rennin enzyme produced by Katy kidneys. This helps to take back more water and salt into the body. They also can have more of the hormone angiotensin which constricts blood vessels in their blood, so that drugs used to lower hypertension are less effective. They are more vulnerable to strokes as are South Asian people. However on the plus side Afro -Caribbean people though tend to have fewer coronary Happy heart problems

Happy heart can stop beating and a person survives, but only if the body temperature is lowered (e.g. falling into cold water). If your body has multiple injuries (trauma) doctors will stop the bleeding and deal with anything major but may leave treatment until the body produces its own defences, otherwise these may not be as effective later after surgery etc.

If you get a cut, us Fred red cells travel to the site (blood flow is increased). Inflammation means that the debris from the injury is being removed through your blood supply. Your Pluto platelets stick together in a plug with protein fibres (fibrin) which becomes a mesh and can form a scab in ten seconds and also release serotonin which restricts blood vessels and blood loss. You may get internal blood leaks also but the body can heal itself. However, dangers can occur where the aorta in the lower half of your body expands with age the walls becoming thinner and then bursts.

There are twelve stages in clotting. Clotting can increase as the viscosity of the blood increases. If one stage goes wrong then the process is stopped. This is a safety net to prevent clotting occurring easily and blocking normal circulation. The Romans, although they did not know about your natural protein fibre nets, used spider's webs coated in vinegar on minor wounds which worked in preventing blood loss. Today you also use plasters to similar effect and sprays to seal up wounds.

With any wound the largest Winston white cells in your blood creep between your tissues to get to the site of injury. They then gobble up Bronwyn bacteria transferred from the object that cut you and those that are already on your Slinky skin. Winston white T cells, on reaching the site pick up the foreign proteins and travel back to lymph node and lymph tissue. They then enrol a whole army of other cells - the Winston's killer T cells, memory and suppressor cells. The T cells also stimulate red Billy bone marrow in which I was born, to produce antigens which will flow back down to the cut to puncture and destroy any invaders. When antibodies destroy Bronwyn bacteria they poke holes in their walls by producing ozone shots which may result in inflammation. Also clotting factors help produce the fibres that trap cells and can help form scabs. . If you get chronic wounds, Bronwyn bacteria can produce biofilms which form layers which cannot be attacked by Winston white cells. Losing a lot of blood can affect Brian brain and you can lose consciousness after ten seconds and eventually can lose kidney function. After losing blood the organs can adjust to a lower blood volume and ration and shunt remaining blood to try and keep some sort of circulation going. Shunting is already common in the body.

Too much loss and Brian brain is affected and the body goes into shock and needs a transfusion. Wounds tend to heal quicker in the desert.

In certain situations like extreme stress some people, as with the horned lizard, can bleed through their Ellie eyes. The horned lizard can squirt blood from both Ellie eyes to over 1.5 metres as a defence mechanism and a repellent. Humans under hypnosis can weep tears of blood and also have Noddy nose bleeds.
Low levels of iron or lack of us Fred red cells in the blood can lead to anaemia. The person becomes tired as less nutrients and oxygen are being circulated. Blood is drawn away from the extremities so there is more passing through Lucky lungs. In extreme cases a person can look ghostly white.

Drugs taken orally, injected or even sprayed on the blood vessels under the tongue also may surround me in the plasma. Pain helps to keep the area still as cells may be replicating. Heat and chemicals produced can cause the pain to be felt. Nellie nerve receptors pick this up and relate this message to Brian brain. Natural endorphins can also help to minimise the pain. However large amounts of pain killer can damage the immune system and you can run the risk of cancer or leukaemia, so you have to be careful. Persistent pain can be tackled by impulse generator electrodes attached to Slinky skin which reduce the pain messages.

Happy heart can get heart disease. This is the most common disease and can lead to heart attacks. Here Happy heart does not receive enough oxygen and the Mollie muscle dies. Heart attacks are often triggered by stress and major traumatic events... Also they may be due to sudden exertion, when Happy heart is made to do extra work and

the blood pressure rises. Football match finals are known
to be one place where heart attacks commonly occur.
During the first three hours of the morning blood vessels
become stickier. Happy heart attacks are common around
10 am but also happen between 8-9 am so try not to exert
yourself during these times. They often seem to occur
more on Monday, possibly due to the start of the new
work week. With age, insoluble pigments can build up in
Happy heart Mollie muscle fibres, affecting contractions.

Taking an afternoon nap for elderly it is said, can reduce
Happy heart disease. Problems can occur when blood
pressure is above 120 over 80 with even higher risk at 140
over 90. Angina can include a pain in the left arm or face
caused by reduced blood flow in Ashley's coronary arteries
(heart disease). Our flow can also be blocked by plaques of
fatty tissue in blood vessels which is atherosclerosis.
Damage to a blood vessel by smoking, bugs and drugs
causes Pluto platelets in the blood to stick to the site of
damage. Then low density cholesterol lipoproteins LDLs,
flow in the blood. As they are small, they stick and help
form plaques. Calcium in my blood then hardens this. The
big white Winston white corpuscles then come along to eat
this foreign tissue causing more damage and bits of clots
break off. These solid lumps float with us can have various
effects including a stroke if they reach Brian brain.
Meanwhile the damage repeats the process making the gap
in the vessel even smaller. LDLs are found in Trans fats
which are vegetable fats changed to solid fat which
increases their shelf life.

Sitting around a long time in coaches, trains or planes, or
after operations can lead to deep vein thrombosis. Pooling
of blood forms clots which can then move into other
organs and stop blood flow. You need to exercise your
feet for three to four minutes by raising and lowering heels
and toes respectively every hour. There are also socks and

stockings similar to those used after operations that can be
used to prevent blood pooling. The term embolism
includes clots anywhere. This is similar to producing scabs
on a cut except this is inside a vessel. Big blockages can
cause loss of circulation and death of tissue resulting in
gangrene and loss of limbs.
Blockage in Happy hearts Ashley's arteries causes Mollie
muscle cell death. Injecting stem cells into Happy heart
through a catheter can regenerate new Happy heart Mollie
muscle and other cells.

Blockage or bleed in Brian brain can cause a stroke,
commonly occurring in people over 65. You may initially
have a mini stroke lasting a few minutes to a few days.
Blockage in the Chelsea Cedric cerebellum in Brian brain
affects motor movements and limb movement. If the clot
instead reaches the cerebral cortex then short term
memory is especially affected or if in the speech areas,
talking. Your speech is slurred and you can't find words,
and are not understood. Symptoms can also include face
paralysis. High level of oestrogen in women from taking
contraceptives or hormone replacement therapy can cause
strokes.

 Both excess drinking and smoking can also increase the
risk of a stroke. Drinking can raise blood pressure
increasing bleeds, while smoking decreases blood vessel
size and can damage their walls encouraging clots.
Sometimes they put chemicals e.g. iodine, into my plasma
to help them see blockages e.g. contrast x-rays or
angiography. For Ashley's coronary arteries they can use
vibratory drills that break up plaque in these arteries.
Reduction in circulation can affect the oxygen content of
the blood. When Happy heart contracts (systole)
Christine's capillaries in your finger or Ewan earlobe
contract and move closer to Slinky skin surface. If light is
shone through these, more light goes through at this time

and is detected. Oxygen content affects the amount of light that passes through. This is made use of in a pulse oximeter which is placed on the patient's fingers.

If Ashley's coronary artery needs a bypass I can get a bumpy ride. A mammary artery from a breast or leg vein may be used to bypass one or several of Happy's blocked heart artery. Also stets (little umbrellas or balloons) can be opened or inflated to increase Ashley's artery size if blocked. When expanded they can squash any fatty deposit present or in the case of using a small umbrella open this so as to hold open the collapsed vessel. Artificial ball valves, donated human valves, animal valves or new valves created from your own stem cells may be used to replace faulty valves.
In some situations a new Happy heart or lucky lung transplant may be needed (pigs Happy hearts can be used) which really sends us on a journey around a Happy heart lung machine. There are also artificial Happy hearts and even small Happy heart like pumps which can be attached inside your leg. In all cases my plasma has antirejection, anticlotting drugs (e.g. rat poison warfarin or other drugs). Also devices can be fixed inside Happy heart which monitor the beat and give shocks if changes occur.

Aspirin, less salt, a low cholesterol diet and weight loss are needed to prevent further problems. Chocolate, it is suggested like aspirin, also prevents blood clots.

We also may get a rough ride and an even greater one when a seven thousand volts defibrillator is used. This is to restart a Happy heart or reset Happy heart's sequence. Also patients may have to have a pacemaker attached with wires to the right atrium and right ventricle. Ultrasound is used to locate where to pass wires through. Today they even have radio controlled pacemakers which are monitored by Wi Fi at a centre outside or at a hospital.

Other Happy heart problems include aneurysms which cause ballooning and bursting of vessel-walls. Inactivity can cause fluid build-up especially in the feet so that foot size may be larger later in the day. Disabled people, coma patients, or the elderly in homes or hospital need to be turned or moved every two hours to get Mollie muscles to force blood through Vernon veins, or pooling and bed sores can occur. Pressure in the body can result in lack of blood and oxygen to tissue. It has been suggested that using sheep Slinky skin with wool on beds would cushion people Often shortage of staff means this is a real problem as Slinky skin and tissue dies and great fist sized cavities can occur. These are especially aggravated where the person has poor blood flow where Happy heart is not pumping properly, often after operations. Also as their Katy kidneys may be less efficient at excreting salt, more salt in the blood may increase their blood pressure.

Bruising is damage to surface blood vessels, blood leaks into tissues and pigments in blood break down from red to yellow and green. Vernon's veins can become varicose. These occur where valves do not prevent backflow and blood builds up giving aching bumps on the leg. These can be dealt with by laser removal and you can use plastic under Slinky skin to prevent re-growth. Haemorrhoids are swollen Vernon veins in Andy anus which itch, and can often bleed, caused by constipation and hard faeces. Athletes competing in warmer countries can tolerate 41 ° centigrade but after a while they can get heat stroke. Heat stroke can affect Happy heart's pattern by affecting enzymes in Brian brain. Sweating stops, the person shivers and becomes delirious. At 26-28 ° centigrade you can die as regulation of Happy heart's pattern of beats is lost. With very cold conditions you can get frost bite which is just like burning. (Clots occur in fingers, ice crystals in the cells). It is also true that you can die of a broken Happy heart. By grieving you may eat more, less the wrong things which, with stress can trigger illness.

Blood problems and immunity

My descendants can be wiped out by radiography. The use of a Billy bone marrow transplant help produce new cells which help to improve immunity. Injection of killer T cells can kill viruses. Stem cells from Billy bone marrow can regenerate organs but this is sometimes dangerous. In tonsillitis and AIDS there are very few Winston white cells so the immune system is lowered

Labelled proteins can be injected into the blood to study disorders. It is also possible to inject fluorescent proteins as was done in pigs, but this means you will glow in the dark if they get to your Slinky skin.

Where a lot of my cells have been lost or are absent due to anaemia. Blood transfusions may be needed

3 IT'S ALL GAS WITH LUCKY LUNGS

Around 1.9 million years ago iron in the earth rusted due to increased oxygen. The presence of oxygen generated by plants led to animal respiration. Oxygen gas is a poison and damages your cells and Diana DNA. Noddy nose warms and moistens the air you breathe. A little sauna for air? Primitive man was very adapted to the cold and had a big nose and nostrils and lots of sinuses (cavities) to warm up the air before it entered Lucky lungs. Four sinuses in your heads make your skull lighter and help to insulate Brian brain from inhaled cold air. Big heaters and big hooters? Noddy noses today can reach one hundred and ninety millimetres. Neanderthal also had big Charlie chests and probably deep voices and there were also many air filled cavities and outlets to equalise pressure.

Today noses can be used for all sorts of things. People can take up milk through their Noddy nose and eject it through their Ellie eye. They can play the harmonica with Noddy nose or recorder through their Ellie eye. They can even whistle through their Ellie eye or blow out a candle and even shoot water through their Ellie eyes for a distance of 3 metres. However you can't sneeze with your Ellie eyes open.

Gills are present in some animals for air exchange. In sea animals these can be many shapes and also on any part of the animal. Once an animal has moved from the sea onto the land the gills need to bet housed in a damp place in the body often with a passage-way a distance away. Diffusion is no good so they need a carrying substance, in our case blood and myself and plasma.

Lucky is a pair of Lucky lungs unlike a snake who only has one. In humans Lucky's left lung is smaller than the right so as to make room for Happy heart. Lucky lung cells are replaced every year but the outer surface of twenty thousand cells are replaced though in a few days, especially in polluted areas. The total area of Lucky lungs is eighty square metres.

Each Lucky lung weighs about six hundred grams. This is heavier than Happy heart (two hundred and eighty grams) and Katy kidney (one hundred and forty grams). Lucky lungs are called Lucky as they are one of the most important parts of the body, next to Happy heart, that keep the body alive.

The air going into Lucky lungs is a mixture of gases. The largest volume being nitrogen. Anything else that is in the air at that time will also enter, dust, pollen, water vapour, tiny seeds, animals, bacteria and pollutants, you name it gets in there.

Mucous forms a layer on Harry hairs in the nostrils and will filter out many of the heavier large particles which will solidify (bogies/snot).

Oxygen on its own is poisonous to the brain, that is why other gases are mixed with it (e.g. helium for divers). Nitrogen also affects the brain as seen in the bends where dangerous nitrogen bubbles form in blood and tissue if divers ascend to the surface too fast. Normally you can breathe out two hundred and thirty millilitres of carbon dioxide per minute about 10 times as much as you breathe in.

Blowing into a paper bag can kill you but can in moderation helps receptors in your body detect the high carbon dioxide and regulate your breathing, stopping a panic attack. Death is often caused not by the lack of oxygen but by the build-up of too much carbon dioxide. An example of the effects of carbon dioxide is given in many science text books. Here a man enters a limestone cave with a dog and only the man leaves. The dog has been killed carbon dioxide which because it is heavier than air formed a layer on the cave floor. Also in the past tramps at night slept next to mounds of freshly made hot lime to keep warm. This gave of carbon dioxide and killed them.

The right lung is larger and therefore takes in more air than the left. They hold approx. 6 litres of air at rest. Approximately eleven thousand litres of air are taken in per day - this could fill a lot of balloons for a big party? .Some people especially seasoned divers can hold their breath for over twenty minutes. They can do this by constricting blood vessels to their extremities, slowing down their happy Heart rate and reducing acidity of their blood. Some are able through repeated buccal pumping to also increase the capacity of their Lucky lungs.

On average breathing in adults is around ten to fifteen times per minute. In babies it is five times faster (baby breaths). You can breathe twenty three thousand times per day. If you live until eighty you have taken approx. six hundred million breaths and breathed in most things I would think! Men tend to breathe slower than women Mollie muscle has a big role to play in breathing as she sits between the ribs allowing them to expand and contract and also has a positive relationship with Dickie diaphragm in Charlie chest cavity.

Dickie is a flexible thin sheet of flexible Mollie muscle under Lucky lungs sealing off the cavity. When Dickie moves down, Mollie's rib muscles move up and out to make a bigger cavity inside. Just like when opening out a plastic bag, air goes in to keep it open. Opening the bag increases the volume inside and the pressure decreases so air at higher pressure outside zooms in to equalise the pressure. This is how air is sucked through your Noddy nose, Charlie chest cavity expands and Lucky lungs chambers expanding. There is fluid in a pleural cavity around Lucky lungs which helps to hold the lining against the thorax wall. You have extra Charlie chest muscles and those in the neck can help respiratory Mollie muscles if breathing is difficult. . Hiccups are a spasm, irritation of Nellie nerves causing contraction in Dickie diaphragm. A fifty year old man had continuous hiccupping after a stroke that's a lot of pardons! Sudden movements cause laughter. Coughing, the releasing of trapped air can force air out at 96.5 Kilometres per hour. Even fish can cough. Sneezing also occurs at high speed. Here Happy heart stops and a chemical high is released giving you a feel good factor.

Once through Noddy nose air travels down through Lincoln larynx (voice box/Adams apple). Your Lincoln larynx has moved to a position lower in your throat than is seen in animals and this could have helped you develop language. Your vocal cords are arranged horizontally inside. They are thicker in men than women and longer the taller you are. Without these you may still be making basic sounds. A reflex that shuts off your airway when eating. Babies do not have this initially and can breathe and swallow together. Pitch is achieved by tightening of vocal chords in your voice box. Loudness is a greater volume of air passing through.
Warmed air then passes down the Trudy trachea bronchus which is a long tube with hard C shaped rings of Candy cartilage to stop it collapsing when you bend your necks. The back of the Trudy trachea has no Candy cartilage otherwise the Trudy trachea would be squashed when you bend over to look at your feet. In Cranes, the bird, Trudy trachea is not straight, but coils first in the body which helps them to make honking sounds.

Eventually in humans air reaches the bronchi twins again having c shaped rings of Candy cartilage for support. One of each of the twin's bronchi tubes goes to each lung and subdivides into lots of smaller tube branches (Brenda bronchioles).

These have no Candy cartilage and lead to very, very thin walled little sacs (Alvin's alveoli) all three hundred thousand million of them! which fill with air. If you could spread them out, the inside area would equal half the size of a tennis court.

I said at the beginning of this book evolution chucked away my Diana DNA but filled me with the iron (my rusty red colour haem) plus protein globin. Oxygenated blood is bright red deoxygenated blood can change to green then yellow in colour. When I die some of the haemoglobin is recycled. My iron haem is great as it can pick up oxygen and help release carbon dioxide. To do this I have to travel very slowly through Christine's tiny capillaries, three hundred thousand of them in Lucky lungs surrounding Alvin's alveoli. Christine's capillary walls are only one cell thick and Alvin alveoli walls are also one cell thick so it's not far for gas to travel otherwise it would take longer (gasp gasp). Blood passes through Lucky lungs once a minute.

The inside of the Alvin alveoli are saturated with water vapour. Here gas exchange occurs across a thin layer of good old water by diffusion (movement from high to low concentrations through the cells). In the foetus a surfactant soapy chemical is produced to mix with the water so as to stop the inside of the Alvin alveolar walls sticking together like old balloons do when not used for a long time. This soap may be absent in some new borns and a chemical has to be injected to help them breathe. Inside the alveoli oxygen is exchanged and also some is breathed out with a lot more carbon dioxide. Nitrogen is breathed in and out of Lucky lungs in similar amounts i.e. little exchange but it does help to inflate and deflate them. Pity really you can't use nitrogen gas it as it would save you eating a lot of meat/pulses (proteins) which you get most of your nitrogen from. Movement of Lucky lungs and Charlie chest is called breathing not to be confused with respiration. Respiration occurs in all cells in the body to gain energy (ATP).

Respiration can be anaerobic (without oxygen) where one molecule of glucose will produce fifty kilocalories of energy or it can be aerobic (using oxygen) which produces six hundred and eight nine kilocalories of energy –quite a bit more!

You have hairy tubes inside Lucky lungs. These tiny Harry hairs (Cecilia cilia) line all the tubes from your throat to the inside of Lucky lungs. Between these Harry hairs are cells which produce slimy mucous which sits on the top of Harry hairs. Harry hairs all move in waves like corn blown by the wind, so that any gunge- dust particles, dirt or Bronwyn bacteria are bought up into the throat and either swallowed or coughed up and spat out.

Breathing is vital for survival and is monitored closely in the body, Chemoreceptors in Brian brain and Lucky lungs monitor the pH in the blood and stretch receptors in Mollie muscles of Lucky lungs register distension. During exercise more Christine capillaries open and increase their diameter to get more oxygen to the tissues so as to produce energy and keep the cells alive.

Lucky Lung related problems

The level of gases in my plasma must remain within certain limits and detectors in the body measure how much carbon dioxide is present. These respond by making you breathe quicker or your Happy heart beat faster or slower.

Serious breathing problems can be dealt with by using ventilation masks or home oxygen kits. Without oxygen being taken in (two hundred and eighty millilitres per minute). Brian brain and cells in the body would lose their energy and die. (Last gasp).

Today asthma is still a major concern. Most asthma is pollen related. In the Mediterranean many children grow up with olive trees and may be affected by the pollen. The smooth Mollie muscle around the airways tightens and the narrow lining becomes inflamed and covered in sticky mucus. Mollie muscle relaxants and even small iron heaters in the airways can be used. Problems are usually larger at 0200-0600 hours. Twenty four days in a salt mine has been suggested as a treatment for asthma. Other lung problems include bronchitis, and inflammation of Bronchi twins tubes can lead to emphysema (fluid in Lucky lungs). Allergies also affect breathing. Medication for lung related disorders involves nebuliser, steroid injections, Mollie muscle relaxers, oxygen, and artificial ventilation

Cystic fibrosis is an inherited disorder. It feels like you are holding your Noddy nose and breathing through a straw. Here Lucky lungs as with other organs, such as the digestive tract, become blocked with thick mucus and the body raises its Fred red cell count to counteract this.

Various organisms entering Lucky lungs can also cause problems. Lucky lungs provide a warm moist environment ideal for microorganisms to grow. The Tuberculosis bacterium (TB) once in the lungs can chop up your thin air sacs and you can start coughing up Lucky lungs with blood.

Psittacosis from parrots and budgies can be a problem as is

bird flu which has been fatal in Asia.

Hiccups are caused by irritation of Dickie diaphragm muscle or an effect on Nellie nerves which causes a sudden spasm. They can in some cases last a long time.

Fungal diseases also cause problems inside Lucky lungs. These can include farmers' lung from animal straw, and thrush from yeasts. Fine dusts are also a great problem and can scarify Lucky lungs and can lead to cancer later in old age. Silicosis in miners is an example as well as asbestosis. The dust from insulation used in lofts can cause problems as can the dust from cutting stone bricks or paving and wood. The correct type of mask is a necessity. Often simple filter masks are used which are not the correct filter size.

At high altitudes breathing and Happy heart rate increases. You can suffer altitude sickness but if you remain here for a long time your body can adapt. More of us Fred red cells are produced and more of Christine's capillary networks are formed. That is why sports people if competing in a sport at these levels often train at high altitudes. Lack of oxygen is like being drunk. Above three thousand metres you become intoxicated. Water taken in to Lucky lungs is referred to as drowning but not always causing death.

Smoking

The bad effects of smoking were recognised as long ago as 1926. Inhaling cigarette smoke produces extra mucous as well as Moby mouth, lip, and throat cancers.

For pregnant mothers the child's birth weight is affected, as the blood vessel diameter is reduced by nicotine making us Fred red cells slow down. This also causes high blood pressure. Smoking is also responsible for asthma, neonatal mortality, less Speedo sperm, stroke and memory loss.

Carbon monoxide in smoke is a hundred times more attractive to my haemoglobin, so less oxygen is taken in. The little, Alvin alveolar cells sacs in Lucky lungs thicken, with more mucous inside so less diffusion of gases occurs. Thick dark tar also collects in the alveoli and with nicotine kills the Cecilia's cilia (tiny Harry hairs stopping them move). These were needed to bring up the gunge you breathe in and out of your Lucky lungs. Bronwyn bacteria now will have a good time in these dark warm tubes. Smoking has been related to eating. Giving up smoking some can cause some to start putting on weight. Smoking can cause emphysema (fluid in Lucky lungs) and also can cause embolism and infection.

Some conditions such as flu or bacterial infections can cause pleurisy which is inflammation of the lung lining. This results in sharp pains across the chest especially when breathing. You are born with bright pink Lucky lungs; the colour comes from us Fred red cells. By living in a city, Lucky lungs become more stiff and fibrous than if you live near the coast. This is due to pollutants. The colour of your lungs also changes. As you age, depending where you live, Lucky lungs get darker and darker. Lucky lung surface is replaced in a few days or longer if pollutants are common. Sending people to fresher air areas (seaside or mountains) was common in the past. This helped clean out Lucky lungs after a lot of coughing. Children born today have cleaner Lucky Lungs than previously as industry has changed but the reverse may be happening in developing countries. or highly populated areas with dense traffic. The incineration of rubbish needs to be closely monitored as dioxins can escape and are well known cancer causing chemicals.

Lung cancer and pneumonia is most common in winter and often comes after flu. Alvin alveoli fill with secretions and Winston white cells. Pneumonia often is the last cause of death for many elderly as other organs begin to malfunction. Bronwyn bacteria can exist in all of you and will multiply when Lucky lungs start to clog up with fluid. Welders also may be prone to pneumonia. An injection is available to provide immunisation for the very young and elderly but this only covers most of the common strains.

4 IT'S A SUGAR BOOST FOR MOLLIE MUSCLE

Without Mollie muscles you could not exist .Your leg Mollie muscles can take us four times around the equator in a lifetime. As your bodies are not protected with a shell or other defensive covering you need to move away from harm. Mollie's Muscles help us move but also help us to generate heat. They also carry electricity like Nellie nerves. Peak Mollie muscle activity these days is at twenty five years of age. Today you have more than six hundred and fifty Mollie muscles attached to your skeleton and this contributes to over forty percent of your body weight- three times as much as your skeleton, quite a lot really.

Caterpillars only have forty thousand Mollie muscles but no skeletons. In water less support is needed by Mollie muscles so animals can become larger e.g. whales. Babies at birth cannot hold their heads up and Mollie muscles for sitting up are developed after 6 months. Examples of your strength include pulling trucks and trains of over one hundred thousand kilograms. Your little finger can support sixty three kilograms but this is a far cry from those who can carry one hundred and ten times their own weight. You need twenty different Mollie muscles for a kiss. Mollie muscles are in a way responsible for protecting the body. This is shown when you wake up suddenly and your entire Mollie muscles contract as your body thinks it is suddenly falling. Possums can play dead for hours, not a single muscle moving, so predators lose interest.

Mollie muscle and friends hold the greatest amount of water in tissues in the body). Water you carry in the plasma adds to Mollie muscle tone.

Mollie Muscles need oxygen for producing energy for movement so us Fred red cells as oxygen carriers, can definitely see to that. More of us travel to the area where greatest activity occurs giving up our oxygen. These are the voluntary Mollie muscles in your arms and legs and also outer body Mollie muscles. Butchers meat is mainly voluntary muscle. The most flexible voluntary Mollie muscle is the tongue which works hard in newsreaders and big talkers. In some animals it has more flexibility e.g. picking up insects in anteaters or in the giraffe which can clean out its own Ewan ears. Some people can read with their tongues as fast as reading with their Ellie eyes. As red blood cells we also pass close to Mollie muscles which carry out the involuntary (without control) movements in your body. Examples of these involuntary Mollie muscles are Gillie gut and the non-stop mover Happy heart or cardiac Mollie muscle.

Mollie muscles get their energy from a substance called ATP (Adenosine Triphosphate) which is the energy currency of the body. ATP derives from glucose which is provided from the blood or converted from stored glycogen in Mollie muscles or Lily liver. Mollie muscle is a much organised protein. When she is in voluntary (striped) Mollie muscle she has two kinds of small fibres or filaments (Archie actin and Monica myosin) arranged in bundles which slide over one another. Archie actin contains many of these filaments and moves inwards and hooks onto Monica myosin so that the total Mollie muscle contracts. For this to happen they need the body's energy packets of ATP and the mineral calcium. To unhook and return (relax) ATP is again needed. If you are dead (last rites for me) then the fibres cannot move back to relax, so rigor mortis (stiffness) occurs. Rigidity starts from your head to your feet, and then the body relaxes from feet to head.

Size wise the biggest Mollie muscle is in the buttocks! (Gluteus Maximus note - the apple bum makes the best race horse. In humans this Mollie muscle cushions you and pulls your legs straight). The smallest example of Mollie muscle is in the Ewan ear only 1 millimetre long. Strength wise in the past Neanderthal man had stronger Mollie muscles, the strongest being in the jaw. They also needed good Mollie muscles to be fit to survive at that time. You still have kept a ninety one kilogram force jaw Mollie muscles for molar crushing but weight to weight the tongue is the strongest. Some people have three tongues. An anaconda can squeeze with a force of four thousand kilograms. Mollie muscles have a lot of power to break or push someone over or even force open a door. This is achieved by putting your whole body into the movement giving extra force and weight. Mollie muscles, if squeezed, could break your Billy bones, but sensors in your bodies within the Mollie muscles prevent this and monitor stretch and tension so that you are not continually laid up in plaster.

Lengthwise the longest muscles are in the thigh. Activity wise the most active Mollie muscles are in Ellie eye which have fine Mollie muscles which can move up to over one hundred thousand times, and can blink six million times per year! (Women blink twice as much as men. I wonder why? -blinking cheek). Mollie muscles attached to the lens are involuntary i.e. you cannot control them. Eyebrows need thirty Mollie muscles to move- (no wonder they can tell us something). Mollie muscles in Ellie eyes can allow them to pop out to over eleven millimetres. Some people can train or have the ability to move their body Mollie muscles in time to music and some can waggle their Ewan ears Mollie muscles can also be protective as with your bum Mollie muscle but also in the ear where Mollie muscles attached to the Ewan ear drum and small Billy bones (stapes) lessen the chance of damage from small vibrations

Endurance training can increase the Mollie muscle and size of the Happy's heart so that the volume increases and Happy's heart rate decreases. Myoglobin, the protein in voluntary Mollie muscle, is responsible for the red colour of your lips (without your lipstick!).

Inside Mollie muscle there are Mollie muscle spindles which have spiral Nellie nerves that indicate how much contraction or relaxation is present. Mollie muscles are strong and also precise for small movements Mollie muscles use oxygen (aerobic) but some can function without it (anaerobic) for a while. Sports people can vary the proportions of each depending on which kind of sport they are involved in. As a Fred red cell I am more involved with the red fast

Mollie muscle which needs more oxygen and contains more Christine capillaries. Training athletes increase the length and thickness of Mollie muscle fibres but they cannot increase the number as that is predetermined genetically. They can also vary the amount of red and white fast Mollie muscles depending on the type of sport involved. A duck is all red meat Mollie muscles while a chicken which cannot fly a lot has white and red Mollie muscle. Lack of training reduces blood flow and the amount of energy producing cell components with increased fat deposits. (The expression is turn to flab). Training every day you need a lot of sugar before and especially after exercise. After the first two hours of exercise any sugar taken initially is converted to glycogen the sugar storage component in Mollie muscle. Moderate exercise will help break down your fat; excessive exercise will affect the breakdown of protein.

Before sport or sudden exertion it is important to exercise slower so as to give the body a chance to monitor and adjust ready for sudden exercise. More of us Fred red cells with oxygen attached and plasma with glucose goes to Mollie muscles so that sugars and oxygen keep them active. If all the oxygen is used up lactic acid crystals build up and ouch! you get cramp. You need to take in more oxygen to remove this. Endurance training as well as increasing heat and lung efficiency can also be influenced by attitude and chemical changes in Brian brain which override the initial pain and allow athletes to reach their limits. Mollie muscles in your right legs are more flexible than in your left so you have a bigger stride.

The other members of the Mollie muscle family are smooth Mollie muscle which have elongated fibres but is not as organised as voluntary Mollie muscle. This is found in Gillie gut. Also smooth Mollie muscle does not tire so quickly (contracts longer) as does voluntary Mollie muscle. (Feed me I'm a good mover) Some Mollie muscles are for gross movement and have few Nellie nerves to stimulate them but other Mollie muscles are used for precision work as in Ellie eye which may have many Nellie nerve cells.

Mollie's family also includes Happy heart with Mollie cardiac muscle. The cells here are arranged like a web and also have small discs along their length which act as relay units for Nellie nerve impulses to travel though the Mollie muscle so that contraction in the top chambers and then the bottom chambers is synchronised. (Let's all beat together?) Happy heart is therefore a super pump, the bottom chamber actually pumping the blood out, the upper chambers collecting the blood until the chambers are full.
Mollie muscle is also important for squeezing blood (us Fred red cells) back up the legs and for shivering to provide energy from contraction to increase body

temperature. You shiver at around 28 ° centigrade. Mollie muscles can generate as much heat as a two kilowatt heater. Body temperature is highest from November to January and lowest from June to July, when the weather could be hot. Lowest values are between 4am and 6 am and highest from 6pm to 8pm at a time when Mollies muscle strength is said to be maximum (tell furniture removers and weight lifters?).

As well as Mollie muscle there are Liddy ligaments which join Billy bone to Mollie muscle (important in movement) –Mollie muscles can only pull Billy bones one way or the other. Liddy ligaments are even found responsible for keeping women's breast the correct shape and also form bands around organs to strengthen them. Some people, due to genetic defects, have very flexible Liddy ligaments and can dislocate their limbs in some cases due to defective collagen (protein) or lack of protein fibres in Mollie muscles/Tommy tendons which also allow them to stretch their Slinky skin which is loose.

There is even an owl man who can turn his head right around and many examples of hands and feet, arms and legs being turned backwards or wrapped around the neck.. Those people who can dislocate their joints can crawl into small boxes or even through tennis rackets! Birds use Tommy tendons at the knee joints and ankles so that they can lock their feet around branches and sleep without falling off .Albatrosses cannot move their wings backwards and also can lock them so that they can soar over the sea for years without expending a lot of energy. Joints and Tommy tendons can also lock in other animals that sleep standing up. Sheep and cows are examples and some can remain in tis relaxed position for months.

Liddy ligaments in the side of the foot are weak which can result in damage by stepping sideways. Tommy tendons

join Billy bone to Billy bone and are important in fixing
and keeping joints together like ropes. The strongest
Tommy tendon is the Achilles in the heel. Fluid filled sacs
(bursae) help prevent friction in between Billy bones and
Tommy tendons. There are several bursae in the bottom
and knees. In the knees the bursae protect the patella when
kneeling
Mollie muscles are fairly resistant to temperature. Naked
you can survive at some very low temperatures if you put
your minds to it (yoga), (several hours at -3 ° centigrade in
fact!) Also you have more chance of surviving drowning
and Brian brain damage if in very, very cold water and
caught within five to ten minutes. The body at very low
temperature shuts down to conserve energy to keep Brian
brain alive. However recovery may still mean that Brian
brain has been affected to some degree.

Some animals can also shut down and even their Happy
hearts stop beating at low temperatures and revive when
warmed up. In hibernation hedgehogs expend little energy
and their Mollie muscles remain contracted and recover
from the front to the back. Some arctic fish have
antifreeze in their blood (these Jenny genes are in some
tomatoes today) -survival strategy. Often body shape
relates to temperature adaptation. Tall Africans are well
adapted to losing body heat while short Inuit Eskimos are
often more a concave shape to keep their body warm.

Living in water all the time you would lose a lot more
energy as your Mollie muscles would have to work harder
than in the air. Man running on land can reach 43.4
kilometres per hour (compare a cheetah 112.6 kilometres
per hour) while in water man can reach just over 8
kilometres per hour (compare a sailfish 110 kilometres per
hour). Most Mollie muscles use oxygen and sugar or
energy. Mollie muscle movement and Lily liver metabolism
affects the temperature. Shivering can raise temperature

and sweating lower it. Energy is even expended when lying in bed (sixty Kilocalories). Walking uses one hundred and sixty kilocalories. Symmetrical women, with good Mollie muscle proportions, appear to produce more female hormones and are more likely to conceive but unfortunately they may be plagued with hormone related disorders later in life. Women have more flexible Mollie muscles than men

Mollie muscle problems

Osteoarthritis is probably the most common problem especially with ageing. This is due to the wear and tear of joints and Candy cartilage on the end of Billy bones being worn down associated with ageing. For hips replacement titanium or stainless steel balls glued into the tibia and plastic sockets in Elvis pelvis can help in this case.
  Rheumatoid arthritis is a far greater problem. It can occur at any age and is inherited or possibly viral. It spreads to all joints. Candy cartilage hardens so joints become immovable and often misshapen. Physiotherapy, anti-inflammatory drugs, steroids cutting Tommy tendons etc. are forms of treatment.

Mollie's big bum muscles pose problems for injections. The needle only hits fat and fluid lingers in the fat. However there is less chance of getting serious wounds if shot as fat cushions bullets. Paralysis of Mollie muscles can be a problem but this can possibly be helped by implanting electrodes into Mollie muscles and relevant Nellie nerves and control movement using radio frequency cuffs and computer signals.
Reattachment of Mollie muscle is becoming an art, a bit of Smartie small intestine has been used as a tongue and limbs attacked by sharks can be reattached. Teflon hollow U shaped tubes can be used to grow stem cells for Billy bone and blood of the owner and growth hormone is also added.

Ageing and muscular systems

Ageing can start from eleven years old. From twenty five to thirty years of age you start to lose your Mollie muscle and replace it with fat. Over forty years old eyesight goes as for example long-sightedness as Mollie muscles weaken. Lack of exercise can cause 0.5-2 percent loss of Mollie muscle per year. Also through ageing the amount of Mollie muscle and Billy bone mass especially in the spine and hips is reduced. With age Mollie muscles and muscular organs get tougher like cooking old meat. Is that why it is said that giants only eat young damsels?

Happy hearts Mollie muscle starts thickening at around thirty years old. Strength size and shape are affected. Over use of Mollie muscles in one's life may make them become weaker. This accounts for older people's jerky movements in arms
Also in women after a number of pregnancies the efficiency of Mollie muscles may be affected. In humans physical endurance has declined by over fifteen percent in the last five years but still some of the over sixties are capable of walking over a thousand miles and doing other energetic activities.

The elderly need to make sure you keep on picking up heavy weights and also push large objects or push against a solid object e.g. a wall and maintain these exercises in order to keep Mollie muscles functioning. Too much time in inactivity can weaken Mollie muscles and too much time in bed can produce bedsores, some of which can led to deep pits and even become gangrenous if blood accumulates and tissues die.
 Also it is important to maintain your inner core temperature. Mollie muscle contraction is important for generating heat. You can lose arms and legs but to survive, the vital organs need to be at the right temperature. In cold conditions you need to keep moving to generate body

heat. If you sit still the temperature of your body falls and this in turn slows down the metabolism reducing energy for further movement. The more you don't move the more your body temperature drops and the less likelihood of moving so hypothermia is inevitable. It is important for the elderly to have heat in at least one room. The hypothalamus in Brian brain is not sensitive to small changes in temperature. Also other disorders that elderly people have may affect their perception of cold or hot. Younger people can withstand very low temperatures In January 2004 a Dutchman; Wim Hof, set a new world record for full body contact with ice. Wearing only his swimming trunks, he stood in an ice filled tube (ice cubes were used) for one hour and seven minutes. He must have moved something to keep warm?

5 BILLY BONE MY BIRTH PLACE

You are born with three hundred Billy bones but end up
with only two hundred and six hollow Billy bones in your
skeleton, which are replaced every ten years. As hollow
they are still very light the skeleton only making up
fourteen percent of the total body weight. Your longest
Billy bone is in the thigh, the femur, forty five centimetres
in men a quarter of a person's height. The smallest Billy
bones are the three in your Ewan ear which are fully
formed at birth, which are used to magnify and transmit
sound The smallest called the stirrup is the size of a grain
of rice.

Prehistoric lizards had skulls over 2.5 metres long,
something of a big head compared to your skulls. Women
tend to have slimmer Billy bones than men with a wider
Elvis pelvis. Also Billy bones in arms are longer in males
than females. Women have their centre of gravity in their
hip men in their upper body. A quarter of all your Billy
bones are found in your feet fifty two Billy bones (ask
footballers).twenty seven in each hand, In the feet the end
Billy bones here are smaller and less mobile than in the
fingers unlike monkeys An elephant on the other hand has
the back half of its foot as fat.

There are still no knee caps in babies but these develop
between 2-6 years of age.
Billy bones normally stop growing at twenty years old. You
doubled your height when you reach two years old. Billy
bone height though can depend on food supply. Your
height increases in the evening and during the night. To
work out your shoe size measure the length from your
wrist to your elbow. Also stretch your arms out at your

side this length is the same as the length of your body. There is also a Billy bone in your tongue (the hyoid) which is not attached to any other Billy bone in the body (good for fast talkers)Your skeleton can support thirty times someone's weight (believe it when you see rows of acrobats on each other's shoulders. A man can balance his whole body on one finger. Pelvic Billy bones transfer weight to the legs for support. Charles V1 of France was convinced he was made of glass and put rods in clothes in case he fell over. Some people have extra fingers and toes. Africans tend to have an extra little finger, but Europeans extra thumbs.

Billy bones offer support and protection, the skull at eighteen months is the size of an adult and is hardened. Wood peckers concentrate their pecking in one area like karate in order to make a hole. The jaw and the skull absorb the shock and there is little fluid in Brian brain so that it is not damaged by the movement.
Billy bones produce us red blood cells in the marrow as well as holding a store of minerals such as calcium. They also are for anchorage of Mollie muscles and Liddy ligaments. Nellie nerves supply Billy bones with pain, temperature and other sensations. Billy bones can feel vibration in the body. Your teeth are the most sensitive especially the canines. Teeth are made of Billy bone. If lost they can be put in milk in a fridge and transplanted successfully within twenty four hours.

As well as travelling to Mollie muscle I often travel back to my birthplace inside Billy bones especially the long ones. Depending on your job Mollie muscles can become more developed and stronger to match your job especially if it involves hard labour. Billy bones can become denser. Tommy tendon joins Mollie muscle to Billy bone and Liddy ligament joins Billy bone to Billy bone. Mollie muscles can only pull, so need to act against each other antagonistically. So when one relaxes the other Mollie muscle contracts to get movement of Billy bones in opposite directions.

Ninety nine percent of the one million animals that exist are smaller than humans and most don't have Billy bones in their backs or spines. Twenty six of Billy bones are in your spine and have surfaces for the attachment of Mollie muscles. Some Billy bones of the spine also fit in one end of the ribs. You have twelve pairs of ribs. Two on each side are floating that is they are not attached to the breast Billy bone in front. Some people though can have eleven or thirteen ribs on each side. One in two hundred women are said to have an extra rib. Why do you have ribs? Snakes use them for walking but in humans the main reason is for protection. Your lower ribs protect Lily liver, Gordon gall bladder Stella stomach and Spiro spleen. The abdomen is not protected by ribs so that you can bend and this allows females to get bigger when pregnant Ribs are also important for lifting Charlie chest cavity for expansion of Lucky lungs. They are also springy and allow for shock absorption.

Your Billy collar bones are separate but in birds and dinosaurs are fused together so as to pull and make a wish, that is the birds. Billy bones in the back, allow the spine to bend. The individual Billy bones fit together with disc pads between. These Candy cartilage discs are very important as they allow flexibility and can absorb a force of several hundred pounds per square inch during exercise. The pads are compressed with your weight during the day and increase in size at night so you shrink in the day and are taller at night. In space you are also taller.

Leonardo da Vinci was the first to show that the spine is curved not straight but forms a flexible S shape. Billy's back bone (of thirty five vertebrae) allowing head movements in the neck vertebrae, rib attachment in the thoracic vertebrae and support for Mollie muscles in the lumbar vertebrae. In the centre of the spine Billy bones (vertebrae) there is a hole allowing Spencer spinal cord to pass through the centre. Regions of the spine are specialised for different functions and the shape of the bones and muscle attachments can differ. There is a pivot system between the spinal vertebra attached to the skull and the next one down which allows you to move your heads sideways. This is like a protruding peg which sits inside a hole allowing side to side movement. To prevent it slipping out is held in place by a ligament. In areas of Burma brass rings are used to extend the neck from childhood. Extra rings are added so that the neck can reach over thirty five centimetres in an adult. If the rings are removed the neck Mollie muscles can no longer support the head so it droops. Some of Billy's spine bones are thicker for support e.g. the Lumbar region,

As mentioned previously, as a Fred red cell, I am created inside Billy bone, especially in the long ones. In babies all Billy bone marrow is red and turns yellow as you grow older. Billy bone can be honeycomb sponge like Billy bone inside the centre and the ends) There are lots of holes for blood vessels to flow through. On the outer side, Billy bone can be compact due to layers like a dartboard with rings of calcium and phosphate. This gives Billie bone strength and lightness. The femur in man has the greatest strength in the body, greater than cast steel weight to weight. Hobbies and jobs can affect Billy bone strength. Stresses and strain set by muscles can increase the amount of Billy bone set. Billy bones adapt to your lifestyle removing or adding calcium to Billy bones into the circulation where it is needed. Horse riders, weight lifters and overweight people may have stronger Billy bones.

In birds Billy bones are even lighter with a thinner outer side that are supported by inside struts. To prevent Billy bones in humans being too brittle and breaking if hit from the side, (a lateral force) running through the rings of calcium and phosphate, (like metal rods in concrete), are protein (collagen fibres). This makes the Billie Billy bones twice as strong as granite. Trees however, which have thickened vessels, can bend even more than Billy bones, whole branches bending in the wind In human bones collagen fibres are linked at the sides by crystals which act like springs to support Billy bones when the weight is downwards and from the side so as to give some flexibility and prevent snapping.

One kind of Billy's bones cells are Billy bone makers or osteoblasts. These get calcium from your diet for storage in Billy bones As a Fred red cell, while on my travels I pass through Billy bones via the blood vessels. The plasma around me carries the minerals, nutrients and oxygen which helps these Billy bone builders make Billy bone. The other kinds of cells are Danny demolition Billy bone breakers, called osteoclasts. These break down Billy bone using very strong hydrochloric acid to dissolve Billy bones and make holes to allow new blood vessels in. Hormones e.g. Patricia parathyroid help to get calcium from Billy bone which acts as a calcium store for body use. Both osteoclasts and osteoblasts are commonly seen after a break. The breakers (osteoclasts) eat away unwanted bits to get Billy bone back to the original size, (well nearly).The osteoblasts, bone makers are positively charged and attracted to negative stress sites. Children's Billy bones mend twice as fast as adult's. Ribs in the foetus if fractured, can be repaired and heal spontaneously and a lot more quickly than in the elderly. Prolonged bed rest or being in space causes loss in Billy bone density. On the subject of fast growth moose antlers have the fastest

growing Billy bone cells in nature. They need to get them
big and lose them later.

Billy's bones link together at joints. The largest in is in the
knee. Fish don't have joints. Your kneecap is not joined
but floats free. Joints in a cow's back legs are arranged the
other way round to yours joints, the knees facing
backwards. In humans when standing straight the knees
lock together so Mollie muscles are needed less.

Cod liver oil benefits painful joints but does not lubricate
them. There are many different joints, a hundred in total in
the body allowing movement. Fibrous joints move very
little and are found joining skull bones. They are also in the
teeth. Cartilaginous joints are able to move under pressure
e.g. between the sternum and ribs. The more moveable
joints are the synovial joints. They have an oily fluid taken
from my plasma (synovial fluid) which lubricates these
joints and feeds the tissues. Your most common of these
can be found in the hinge joint at the elbow and knee.
Apes' front limbs do not straighten as they have no
elbows.

Other synovial joints include the ball and socket joints
which are the most mobile and found in the hip and
shoulder. As they are the most mobile they can come out
of their sockets fairly easily. Some people can dislocate
them in order to perform feats and can crawl into small
spaces. The hip joint is the strongest.
There are also joints that work like a saddle as well as
sliding ones. The smallest joint is in the Ewan ear. Women
have more flexible joints than men. Humans also
developed a chin to support their jaws and this helped you
with your language. (chin chat?). Joint capsules have
position receptors.

Human joints and other animals are held in place by Liddy ligaments, while Tommy tendons join the Mollie muscles to one of Billy bones to make them move.

Billy bone problems, disorders and effects of ageing
Posture is most important to prevent back problems. Sitting up straight has been shown not to be the ideal posture. This is because all the body weight is concentrated on the lower section of the spine
Your Billy bone cells normally will last ten to thirty years. By the age of thirty your Billy bones begin to shrink. Your ratio of height to the distance from your belly button to the tip of your feet is always the same i.e. 1.618 (example of Fibonacci numbers) try it.
In very rare cases some people can have two Elvis pelvises and four legs! Some people can have their knees round the wrong way. Also some children are born as sirens/mermaids where their lower appendages are fused together.
It is important not to lose patience with clumsy teenagers and elderly people who can appear to be clumsy. The reason they are clumsy is that teenager's Billy bones are growing faster than the Mollie muscles so they have sudden movements and for the elderly Billy bones are shrinking and also Mollie muscles are weaker so the ratios are upset again.
From the age of thirty five the spine shows wear and tear. Movement helps pump liquid into discs between vertebrae. High heels affect the spine curve the toes and cause arthritis. The disc pads between Billy bones can also shrink with ageing. These can be bolted together and a plaster solution placed between. In some cases stem cells can be injected which have been obtained from Billy bone marrow chemicals and collagen can also be added and the new discs can last over thirty years it is said.

At menopause the female hormone oestrogen drastically falls, resulting in loss of calcium. The female hormone oestrogen helps calcium to be taken up normally. Osteoporosis can occur. Here there are more Billy bone breakers dissolving Billy bone than Billy bone makers. This means that Billy bones become weak with holes in them just like a crunchy sweet and can cause fractures at wrists, Elvis pelvis etc. For overweight people this is especially dangerous where their hips can crumble. To combat this, strength building exercises are needed and hormone replacement therapy.

If vertebrae start to thin then they can collapse (Compression fracture) and the spine can be bent over .This is known as Dowager's hump and you can see little old ladies bent over double due to this. It is essential that you eat lots of calcium rich foods and drink plenty of milk as a child to build up your calcium reserves. Also for women when pregnant, to take extra calcium where your child is taking up your supply needed to build bones. In old age your moving parts are affected first. Prolonged exercise for anybody at any age can be damaging leading to osteoarthritis.

Osteoarthritis is the wear and tear of your joints. Candy cartilage is worn away and Billy bone rubs against Billy bone. Cherries and ginger are good for bringing down inflammation in joints and it is believed chondroitin sulphate helps with Candy cartilage production. Artificial joints can be made, especially ball and socket as hip replacements. Here a stainless steel or titanium or chrome ball (iron would rust) with a spike is hammered or screwed and glued into the sawn off end of Billy bone femur and a plastic or other material socket fitted into Elvis pelvis or shoulder. Where bone has been removed because of cancer or other causes, a piece of bone from the hip or leg

can be used to replace it. If not stem cells can be grown on
a mesh to produce an exact replica for replacement.
 Bed rest weakens Billy bones and Mollie muscles causing
more falls and fractures when moving around. It has been
suggested that older people do Tai Chai as this can
improve balance. Old people also get rheumatoid arthritis
where Billy bone outgrowths occur which seal up joints
and prevent movement. It is important for the elderly not
to slouch and remain upright.

Paget's disease causes old Billy bone to be replaced by
weaker Billy bone and can cause deformities and Mollie
muscle problems. There is even a syndrome called Stone
Syndrome. This is genetic and Mollie muscle can turn into
to Billy bone producing a second skeleton gradually
turning the patient into a solid mass of Billy bone like a
stone.

Rickets are deformed Billy bones often seen as curved
limbs due to lack of calcium from Vitamin D deficiency.
Nationalities that cover their bodies and faces up when
outside, can often develop rickets. Sunlight helps build up
vitamin D through Slinky skin.

 In Victorian times some women had broken ribs which
was due to wearing very tight corsets. Even today shoes
can have important effects on Mollie muscles and even
Billy bone so it is important to wear the right ones.
Caffeine from drinking coffee is believed to lead to
leaching of calcium from Billy bone
Brittle Billy bones can be a genetic disorder and just
reaching out or sneezing can in extreme cases break a Billy
bone. Even in Wendy womb the baby can get bone
damage where children have this disorder.
To have a Noddy nose change you need to break Billy
bones with a hammer. On stage never say good luck but
break a leg, Billy bones will remain are useful for making

gelatine and to thicken foods and also for the soil as bone meal, a source of calcium and phosphate.

6 SLINKY SKIN, CANDY CARTILAGE AND HARRY HAIR

You can't beat me as I am Slinky skin, the biggest organ in your body. If you measure from the sole of your feet to your navel you get the golden ratio - PHI which is 1.618, Your slinky skin weighs five kilograms and you have 2.3 square metres. of Slinky skin which would cover a single bed. The outermost layer of Slinky skin is the epidermis which divides, leaving dead cells on the outside which can harden.
Slinky skin cell life is around two to four weeks and can result in over thirty layers of dead cells on your Slinky skin surface. You can lose fifty million Slinky skin cells in a day and you get one hundred new Slinky skins in a lifetime – just over 1.8 kilograms! I just changed mine last week? - Or I just jumped right out of my 99th Slinky skin. Dead Slinky skin these days can be removed by defoliation using Japanese silk pulled across Slinky skin.

The ancient Romans and Greeks used strigils which are curved metal scrapers, to remove dead Slinky skin and dirt. The Georgians also used similar items to scrape their tongues and these are still available today. They put oil on Slinky skin then scraped. The next layer in Slinky skin is the dermis. This contains sweat and sebaceous oil secreting Gladys glands. Slinky skin secretions act as an exterior defence method with their acid secretions. Most of Gladys's sweat glands in humans are found on the hands, not feet. Palms sweat the most. Sweat can be blue, yellow or green and is found around the armpits, genitals and breasts. Eating a curry revives Gladys's sweat glands. Babies do not have Gladys's sweat glands when very young but these develop later. Therefore it is important not to let babies get overheated. Dogs and most birds also do not have Gladys's sweat glands but can lose water by panting. In men sebum from Gladys's sebaceous glands is more common on the face than for women. At puberty with an increase in sex hormones, sebum when exposed to air darkens and can make fatty acids which Bronwyn bacteria love. White blood cells then invade the glands and sometimes pus is formed from dead bacteria leaving black heads. People often squeeze these out in long lengths, hence their nickname worm pimples. In the Ewan ear you have over two thousand Gladys's wax glands but less is secreted as you get older. This has a smell repellent to insects and also picks up any dirt, dust or micro-organisms entering the ear.

Also many other tissues are present in the dermis. These include Mollie muscle, Nellie nerves and connective tissue. Throughout this layer are Christine capillaries in which as a Fred red cell I travel through. Two and a half centimetres of Slinky skin can contain over 6 metres of blood vessels. Some people can pull their neck Slinky skin over their Moby mouth. The next layer beneath the dermis is a fat layer which can vary in thickness depending on what sex you are and where in the body this is found.
Slinky skin can vary in thickness from 1-3 millimetres in places. It can also be as much as twenty layers thick. .A hippo has Slinky skin thirty eight millimetres thick. The foetus has fingerprints at three months. Some cats have wings of Slinky skin covered in Harry hair and Mollie muscle. But unfortunately they still can't fly.

Your Slinky skin also has numerous sense organs which have direct links with Brian brain. Glands in the skin around the eye produce tears. These are ninety eight percent water. By crying you can produce sixty five litres of tears in a lifetime. Tears as with sweat contain the enzyme lysozyme which kills Bronwyn bacteria. Women's tear glands are more active and they cry more. Gorillas do not cry and some animals, except elephants do not seem to mourn the death of a partner or offspring. Slinky skin may also contain pigment cells or melanocytes containing Melanin. More pigmentation can occur as you grow older and pigment cause spotting on back of hands. Polar bear Slinky skin is black; their Harry hair is not white but clear. Sweat and tears contain lysozyme which kills Bronwyn bacteria. Transfer of heat is through the hollow Harry hairs.

Slinky skin problems and adaptations

Beauty is said to attract a mate and indicates health .The more attractive the face depends on the ratio of the width of Noddy nose to the width of Moby mouth. Marilyn Monroe or Queen Nefertiti are examples where this ideal ratio is seen. Dimples, small depressions, are considered attractive and can occur where Mollie muscle is attached to Slinky skin, seen in the cheeks sometimes. Dimpled Slinky skin can indicate cellulite which is ripples of fat that get trapped between the fibrous bands that connect Slinky skin's tissues

 Slinky skin starts ageing at twenty. With age your connective tissue loses its elasticity so that Slinky skin sags e.g. bags under Ellie eyes. At around thirty your Slinky skin starts to get thinner as more Slinky skin cells are lost and the elasticity of Liddy ligaments is getting less. Your Slinky skin does not go back to its normal shape and give wrinkles at the age of fifty. This is accelerated by smoking and the sun. At fifty you may also start to bruise more easily.A beauty treatment now exists previously used to treat sports injuries. You enter a chamber at 120 ° centigrade for a few minutes which is believed to improve Slinky skin. Laplanders covered Slinky skin with rancid fish oil to endure very low temperatures. Some grease themselves. Some people never wash and may never have washed for over sixty. Their Slinky skin is dark and scaly. They may also wait for their clothes to rot and fall off. In some parts of the world humans have really thick callus on feet and soles so can walk on rocks You can get a scab on injured Slinky skin in ten to fifteen seconds but this may leave a scar unlike Billy bone which heals without scars. A Chinese man was born with transparent flesh so you can see his Billy bones and organs. This is similar to the banana tree frog whose Slinky skin is transparent so you can see their Billy bones. There are many examples where people threatened with execution sweated blood from every portion of the body. You can even get bloody tears if under extreme stress. In the late evening Slinky skin sensitivity increases Itching cells release histamine which affects the Nellie nerves making you want to scratch.

Candy cartilage grows in Ewan ears and Noddy noses of older people. Some children when born can have extra Ewan ears on their neck. In China long Ewan ears are considered lucky. During the 10th century in China some children had their feet wrapped tightly and rewrapped each day. This resulted in Billy bones breaking and their toes becoming bent. The children's feet sometimes only reached three and a half inches.
Leeches suck out blood and are used today to prevent bruising of Slinky skin especially Ellie eyes. They are also used to draw blood to the end of the finger when the tip is reattached to encourage blood flow into the attached tip. Heat stroke causes headache and fatigue but also stop sweating and then you can lose consciousness. Drinking too much water can cause sweating to cease and you can become helpless Fat can be for insulation and help people float and survive better in water than those with less. At times it has also proved useful in stopping bullets. Spontaneous combustion is believed to be due to ignition

of fat stores. Fat burns. Nero put Christians in his garden as human torches.

Cutaneous porphyria causes darkening of Slinky skin and reddening of teeth. You could be considered to be a werewolf. Urine can also be pink, brown or purple as a result of. Hypohydrotic ectodermal dysplasia also produces a Dracula like appearance. Humans only have with a few pointed teeth. They also cannot sweat and have to be kept out of sunlight otherwise they would overheat.
On a bed of nails penetration of the skin depends on pressure. As long as the number of nails present is proportional to the body weight of the person the combined pressure on each nail is not enough to penetrate the skin.

Some people get crocodile like Slinky skin which is scaly grey like a lizard also seen in new born babies. With severe psoriasis you can develop a leathery skin over the body. Tree men exist where a genetic disorder results in large outgrowths, huge wart like branches occurring on the hands and parts of the body become hard like a coral reef. In some other disorders moles can grow to cover the whole of the back. Elephantiasis produces thick Slinky skin on humans and swollen limbs like an elephant. like an elephant and is caused by transmission of parasite by a mosquito.

The elephant man had Proteus syndrome where Billy bones and Slinky skin had tumours. Noddy nose and lips and forehead projected like an elephant, papillae were found on the back band other on places. Hypertrophy of some Billy bones was present and he could not hold his head up. Some people may also have macrodactyly where their fingers can be as thick as arms and their whole hand bigger than a football.

Tumours can weigh a lot 137.6 kilograms and reach one metre across.
A third of all cancers are sun related. Sun tan lotion of 40-50+ should be used on white children in very hot climates whereas black/Asian should be around 35 and this repeated every twenty minutes. Nobody should be exposed to high radiation from the sun especially at midday for more than twenty minutes at a time, Cells will then begin to cook, Babies under 6 months should not be exposed to sun light at any time and children should not be out at 11am-3 pm when the sun is hottest. Though cancer can be inherited it may not be triggered. Melanin is produced in sunlight is protective. in Slinky skin giving you darker freckles. Ultraviolet light can even cause tan on a cloudy day as radiation can still pass through.

Sunbeds using UV on younger Slinky skin increase freckles, moles and Lily liver spots which can turn cancerous in later life. Normally your Jenny genes can repair damage or cells can die or not reproduce. Mobius syndrome is where there is no expression on face as paralysed. and you cannot move your Ellie eyes side to side. Slinky skin can become affected by Bronwyn bacterial and fungal disorders. Louis X1V did not look after his feet so that near his death, parts of his feet were so bad that his toes rotted and fell off in his socks. Henry the VIII also had gangrenous legs which smelled.

There are many cultural trends and Slinky skin_has now become a canvas for art. Tongues can be split into two like those in snakes. snakes. Wooden discs placed in the lips of Amazonian Indians are considered a status symbol. They have Different sizes of plates, are used the size increases with stretching. These large soup plates can reach thirty centimetres in diameter in the bottom lip. They can be removed at night and when not in company. Some people stretch Ewan ears using stones as weights. Branding is also common, African tribes have scars made on faces and ash is put onto wounds to prevent them closing. This even occurs in European countries where wounds are kept open after cuts to ensure scars are produced.

Piercings can be traced back over five thousand years, and seen in the Aztecs, Eskimo and Native Americans. Nipples were pierced as a sign of bravery in Roman times. Today piercings can be on any part of the body. Tattoos also are common place and increasingly becoming common in women.

Shrunken heads involves removing the skull and placing hot water sand or stones inside until shrunken then moulded into shape and stitched. Henry VIII, if he found a poisoner would boil them alive. Today plastic surgery has reached the state of the art. Limbs and tongues can be transplanted but need to overcome the immune responses. You can now have a face transplant etc. produced by a 3D printer to be exactly like your original features. Slinky skin is useful as a food an example being the facial features of animals and tails being used to make sausages.

Slinky skin is useful for detecting allergies by pricking or coating Slinky skin with a substance or substances which are common allergens. Some dyes in Harry hair dyes produce an allergic response and are tested on Slinky skin behind Ewan ears. Some rare allergic reactions to some antibiotics can cause sloughing off of the surface of Slinky skin and blisters all over Slinky skin, with severe salt and fluid loss. Patients may need a Slinky skin substitute to recover. Another allergic reaction can cause all the fat under Slinky skin of the face to be removed so that the person looks fifty years older (lipodystrophy). Another unusual allergy is to water called aquagenic urticarial. Here the sweat or tears can cause a reaction and people have to avoid the rain or suffer pain during a bath or washing.

Harry hair and Nelly nails

You are not as hairy as your ancestors but Harry hair roots occur all over your body but Harry hair is not pronounced or as long as monkeys. Goose pimples occur where Harry hairs would have been and sites become erect where Harry hair would have kept you warm. Pubic Harry hair remains it is believed, as it once was used to attract the opposite sex, as well as helping to evaporate pheromone scent given off.

Harry hair (all five hundred million of them on your body except your lips) is made out of the protein (keratin) the same stuff as Nora nails. One Harry hair can support 3 kilograms and for thickness is comparatively stronger than steel. You have the same number of Harry hairs on your body as a chimpanzee (though not as long) but they are a lighter colour. You can have huge clumps of Harry hair on your bodies; some parts of the body are hairier. Your eyelashes are also keratin which like Harry hair, is replaced but lasts for longer (estimated you lose thirty metres of eyelashes in a lifetime.

When it is visible it is dead, growth occurring from the base. For Harry hair there are three stages three to four years of growth followed by a resting stage for a few months after which it falls out. Growth on head Harry hair is faster than Nora nails. Harry hair normally grows at 1.25 centimetres every month. When the weather is hot Harry hair grows faster. Beards are the fastest-growing Harry hair on the human body and can grow to nine metres. Pubic Harry hair can grow to over seventy centimetres. Nora nails grow quicker as you age. The longest as a toe nail can reach 8.5 metres Horses nails have to be cut and horse shoes are nailed on to protect them from cracking on the hard roads. They say that Harry hair and Nora nails continue to grow when you're dead. This is untrue. They may look longer because of tissue shrinkage. Harry hairs cannot grow on the areas where you have been vaccinated as Harry hair roots have been damaged. Wet Harry hair is over one and a half times as long as dry. Blondes have thinner Harry hairs than red Harry haired people and more Harry hair than black Harry haired people which is probably due to more of the female hormone oestrogen.

Hairy eyebrows exist to deflect sweat down over the cheeks instead of into Ellie eye which can affect vision. In the 17th Century women added extra Harry hair to increase Harry hair height. They often used horse hair frames to support this Harry hair and added flowers and even stuffed birds. Unfortunately mice sometimes made nests inside them.

Harry hair problems
Harry hair has an outer cuticle like layer which can be damaged. Healthy Harry hair can stretch to over one hundred and fifty times its original length. Damage can be caused by over styling, too much washing, the shampoo you use, over brushing, too much heat when blow-drying, These factors can also cause split ends where the cuticle is damaged at the end. Split ends cannot be repaired only cut off or substances used to stick Harry hairs together. Over one kilogram of Harry hair can be lost each year. Men and women can lose Harry hair in certain regions, men in the front and top and women the top. This is often related to the genetics of your parents Physical or emotional stress can cause Harry hair loss.

 Of the one hundred thousand Harry hairs on your head, these can start dying losing their pigment so that they turn grey but this may be genetic not ageing. Sudden shock can have the same effect and can even turn grey overnight. Some people especially during puberty produce more of the natural waxy oil sebum from Gladys sebaceous glands which makes their Harry hair very greasy. Some people

pull out Harry hair eyebrows, Ellie eyelashes and are unaware. This is called Trichmaia.
In Harry hair follicles one can find mite Desmodex who lives lays Elsie eggs at the base of Ellie eyelashes. You can also get head lice and ringworm and psoriasis in some cases.

7 GILLIE GUT MEETS VICTOR VILLI WITH A LITTLE HELP FROM FRIENDS PETER PANCRES AND LILY LIVER

As the climate became colder your ancestors had to store food. All foods are for providing energy for bodily functions, even licking a stamp uses one calorie. . In some countries people eat high proportions of one item of food to obtain energy Also food provides macro and micronutrients. The macronutrients are large building block molecules, proteins, fats and sugars while micronutrients include tiny amounts of minerals and vitamins used in enzymes, transport etc. Thirst and hunger are due to changes in the chemistry and are detected in Brian brain.

Temperature and enzymes are important for digestion. Some birds and lizards sit in the sun to help increase their temperature to aid digestion. You don't have to travel far to find food but an albatross can travel over one and a half thousand kilometres to get food, flying over the oceans for several years. You also have a wide variety of food sources. In the past cannibalism was common you ate your enemy or parts of them. Victorians however ate things which you often don't eat today e.g. battered mice and stewed bluebottles.

Processed food today is still not always pure and can contain insect legs and even rodent droppings. In a life time you are supposed to have eaten at least eight spiders. Limits are set for the number of insects, maggots, and mice droppings allowed in food by weight. Female black widow spiders can eat twenty males in a day. You can eat as much as 6 elephants in a lifetime while a shrew eats its own weight in thirty six hours. An octopus can eat themselves if upset. Certain foods are repellent to humans and are poisonous even contact is poisonous.

Now I am off for a journey through Gillie gut which is replaced every six days. Food takes twenty four hours or even up to three weeks to digest through if fibre is present At six weeks inside your mother you are able to swallow, absorb and discharge fluids. At eight weeks your Moby mouth opens and closes. You all get your energy from the sun using photosynthesis, which stores it in plants as large organic compounds in chemical bonds long, long, long chains which in turn are eaten by animals. You eat the plants and/ or animals to break down these long, long chains to get at this energy and to make your own long chains. Monkeys, except for tarsiers, eat leaves fruit etc. and in some cases chimps are carnivorous. Carnivores don't eat food struck by lightning (one million volts a flash!).Possibly this affects the taste.

You rely on enzymes and have a lot more enzymes than living plants. You can survive thirty days on sugar and water. Sugar in the 14th century was only used as a medicine. Each time one thing eats another you unfortunately lose ten percent of the energy in the food therefore it is more efficient to eat plants than animals which are higher up the food chain. That's why thousands of people can be supported by eating rice, wheat and corn. China produces over thirty two percent of the world's rice. Food also provides a source of water in the food itself or from its breakdown. (1 gram of fat oxidised gives 1 millilitre of water, while one gram of sugar gives 0.6 millilitres and protein 0.4 millilitres of water. Some foods use more calories to break it down than is actually in the food e.g. celery. You also get chemicals, minerals and additives from your surrounding earth and in water e.g. salt. As a Fred red cell I have a big role in Gillie gut. I supply all the intestinal cells with wonderful oxygen and my surrounding plasma picks up food that has been broken down to small molecules and carries it around the body to organs, tissues and cells which are waiting for it.

People can distinguish between three thousand and ten thousand different odours. Smell initially is more important as there are more sensory receptors for this than taste
All food is checked by the senses before it enters Gillie gut which is a nine metre long tube and just an extension of your Slinky skin. Your Slinky skin goes in your Moby mouth and out of your Andy anus i.e. Gillie gut is not really inside your body (who said your Moby mouth looked like a mirror image of your Andy anus). Enlarged cavities in Gillie gut hold the food and regulate its journey. Mollie muscles help to move and mix the contents. Valves in your Gillie gut regulate food as do Mollie muscles and other organs link to it with tubes to inject fluids to help breakdown food. Nellie Nerves and hormones control Gillie gut and a blood system runs close to it so as to remove required fluids and salts for use in the body.

Your tongue is the strongest Mollie muscle in the body and helps to mould food and put it to the back of Moby mouth. Your tongue is fixed towards the back of your Moby mouth unlike frogs where it is fixed in the front and can be shot out and flipped back. Your tongue can also be a long as 9.5 centimetres from lips to back end. Moby mouth contains ten thousand taste buds which are more sensitive in the morning and a thousand flavours can be distinguished. Fat has flavour and is a feel good one without fat there is little flavour. Receptors for taste are also on the insides of the cheeks and palate. The tip of your tongue detects sweetness, the sides and front bitter and salty and the very back sour. Cats, it is believed, cannot taste sweet. Butterflies and other insects taste with their feet. Humans can taste one gram of salt in five hundred millilitres of water. Detecting salt is due to more sodium entering the Nellie nerve cells, bitter and sweet is due to special proteins which pick this up and sour is due to more electrically charged hydrogen atoms being present (like an acid). The type of water added to drinks i.e. soft or hard can affect the acid or alkaline taste of drinks such as tea or coffee.

Every ten to twelve days ten percent of taste buds commit suicide so that fresh ones take over. Girls have more taste buds than boys. More buds are also lost as you reach your seventies so give your granddad extra spicy food. The arrangements of your taste buds etc. gives you all a different pattern like finger prints. So instead of using fingerprints or Ellie eyes (iris patterns) perhaps you should pull your tongues out like New Zealander greeters. Also your Noddy nose helps taste, as seen when you have a blocked Noddy nose and have poorer taste

Our tongues are also usually pink. Different colours over the surface of your tongue can show Chinese tongue readers which part of your body is upset or ill. Whole books exist on tongue diagnosis. A purple tongue relates to blood and can indicate poor circulation. Cracks and marks also have meanings. This does not work though with black bears as they have a blue tongue! Perhaps there is another atlas for a bear, who knows. Your tongues are soft but a giraffe's is longer and muscular and its lips are like leather allowing it to eat leaves on spiky branches and lick its Ewan ears. You can only lick our Noddy noses, Llama's tongues are short so they can't push out and lick. Crocodiles also cannot stick their tongues out.
Drinking sugary drinks prepares Gillie gut for a meal. Just as Slinky skin scales are lost (house dust) so are the cells

from the lining of your Gillie gut. Your Gillie gut moves just like a worm. Circles of Mollie muscle squeeze the food forward (peristalsis), whereas longitudinal Mollie muscles in front of the oncoming food shorten and expand sections. Mollie's Gillie's gut movement is minimum between 0000 hours-0200 hours and maximum between 2000-2200 hours. Bowel movement (lower end of your Gillie gut) is maximum between 0800-1000 hours and between 2200-2400 hours.

Imagine a chip has just arrived in Moby mouth only to be instantly cut by your front teeth the incisors (flat and chisel likes). This is then torn apart by canines (sharp and pointed your fangs) and the chip is ground by premolars and molar back teeth. Grass feeding animals and elephants depend on these. Toads also have teeth, but poor frogs don't. Male horses have forty teeth but females have thirty six. Slugs/snails can have over twenty seven thousand teeth but this is on a sort of conveyer belt of teeth which rasp away at the vegetation they eat. Crocodiles cannot bite bits off so revolve to tear bits off their prey. Some babies are born with one or two teeth but these are loose. Some children can have a row of adult teeth behind their baby teeth. For big teeth elephant front teeth can grow 3 metres.

 For strength and wear and tear, rats have strong teeth and can bite though thick metal and wire cables. In humans sitting in the gum are twenty milk teeth waiting to come through which they do at about one to two years of age. At 6 years old thirty two teeth have grown. Sixteen at the top and sixteen at the bottom. They start losing these this age and girls tend to lose these later than boys. Some elderly people can retain all their first teeth. or may even have had three or four sets of teeth. Lemon sharks grow a new set of teeth every 2 weeks. In humans a tooth is glued

in place and can support sixty four kilograms. Teeth were removed from soldiers in battle to make dentures.
Eating cheese and drinking milk provides calcium and phosphate which combines with Sally saliva and helps prevent erosion of enamel on the teeth. Piranhas have very sharp teeth and in schools can strip you to a skeleton.in 3 minutes.
Unlike cats and crocodiles you can move your jaws side to side as well as up and down. Apes, like primitive man, have a strong jaw. Mollie muscles help them to crack up Billy bones to get at the marrow. This meant that Neanderthals' teeth were often worn down. Human jaw Mollie muscles are still strong today. Some people still chew on Billy bones, crack open nuts and pull off drink tops from bottles with their teeth but this is still a great way from what you used to be able to do. Incisors, your front teeth, have a weaker force (max two hundred Newtons) than the back molars which can exert a lot more (nine hundred Newtons). Hyenas can exert eight hundred kilograms/squared to crunch Billy bones. Some snakes have four rows of teeth which move over prey to drag it into Moby mouth. Cockroaches have teeth in their Stella stomachs.

After tooth attack the chips are attacked by Sally saliva from the Sally salivary Gladys gland family (she has 3 sets). She has the same salt content as my plasma and can, when in full flow, produce fifty thousand litres in a lifetime. This is roughly 2 litres per day, two swimming pools full but who wants to swim in Sally saliva? The average talker sprays three hundred drops per minute so avoid loud and long speakers. Sally's saliva still can't keep up with eating crackers as fewer than ten can be eaten in a minute. Sally's salivary secretions can be sticky or watery and contain the enzyme amylase (remember this by using lazy Amy as a reminder) to break down long starch (e.g. rice and potatoes) molecules to smaller sugar molecules. Spiders inject their Sally saliva into their prey and flies smear it over the food then suck up the result. Good job humans don't do this. I've not heard of a human who can inject Sally saliva but some certainly do cover food with Sally saliva (lolly suckers). Sally saliva also contains the enzyme lysozyme which is also found in tears and pops and kills Bronwyn bacteria.

It is believed that you, like some animals and birds, in the past chewed your food and then passed it onto the baby. This is still done in some cultures today. This, some say, could be the origin of the kiss! Sally saliva can react with fillings to form semiconductors and they say anything that vibrates acts as a speaker in your head allowing you to hear music.

Back to Gillie gut. The uvula at the back of your throat hangs down and swings. This helps swallowing and stops food going up your Noddy nose. The chip which you ate is directed down the food tube where epiglottis (a flap) prevents food going into the windpipe). In large snakes the end of the windpipe dangles out of Moby mouth so that it can still breathe while swallowing large prey. The food tube called Owen oesophagus is twenty five centimetres long and 2.54 centimetres wide. It takes about seven seconds in humans for food to reach the entrance to Stella stomach. It then passes through a valve (which is actually part of Dickie diaphragm) through which the tube passes into the acid bath in Stella stomach. Owen oesophagus collapses when there is no food in it.

Some animals, such as cattle, sheep and llamas eat cellulose. Cows have more than one Stella stomach to help with digestion. This is helped by Bronwyn bacteria. Some monkeys eat leaves and these have ruminant like Stella stomachs. Even some birds may use the same method and have two Stella stomachs. Pandas can eat over 4 kilograms of bamboo shoots and leaves in a day. The fibre in plants though is not digested as in horses and can be seen in their faeces. Asians also eat more fibre than people from some other countries. Some birds, like reptiles and dinosaurs, had stones in their Stella stomach to help grind down food. Just as some pet birds need grit to help them do this today. Cows have Bronwyn bacteria in their Iris intestines which can make protein.

Stella stomach can be the size of a fist when empty but expand to hold a maximum of four litres and produce 2-3 litres of gastric juice per day. The presence of food stimulates chemoreceptor and mechanoreceptors. Stella stomach can expand by several centimetres when she begins to fill up and she gets smaller and folds in with gradual food loss from your Andy anus. Starfish can turn their Stella stomachs inside out to get inside shell fish. Toads can regurgitate their Stella stomach and then suck it back. In humans Stella has Mollie muscles in three directions so is a good churner. It can take seven hours to fully break down a meal.

Abdominal Mollie muscles can also help squeeze Stella stomach. This is especially noticeable in small children. She produces really, really strong hydrochloric acid PH 1-2. This could burn a hole in your carpet if we cut you open. Eating meat, drinking alcohol or coffee produces gastrin, a hormone in Stella stomach which starts the release of hydrochloric acid and churning movements. She also wisely produces mucus (replaced every two weeks) to cover her inside lining so that the acid does not burn a hole in her walls (Stella stomach ulcers?). There is less of this mucus wall cover in the elderly so more ulcers occur. Stella's stomach acid kills any bugs that enter with the food as well as removing the water from food to break it down. Unfortunately drugs can be broken down by the acid in Stella stomach. More acid is present in the night than in the day, with a rise at 10pm and 2am. In some cereals iron filings are included (can test with a magnet). These are broken down by the acid to give humans an extra source of iron. Add a splash of orange juice and more iron will be absorbed by the body.

Any protein in food is attacked by Stella stomach. The enzyme Polly pepsin which only works in Stella stomach and acid conditions does this. Pepsi cola also is believed to contain this enzyme. Note this is not the same as coca cola; this was developed as a Brian brain tonic initially! Only the amount of protein needed is used by the body. Extra protein is excreted.

A relative of Polly pepsin (a sort of pre-enzyme, (the body has many) is found inside the cells of Stella stomach. This pre enzyme only turns into the enzyme when it leaves the cells to enter Stella stomach cavity. If Polly pepsin enzyme which attacks proteins was inside the cells of Stella stomach wall you would digest your own Stella stomach! Hey where's my Stella stomach gone? Stella's digestive capabilities were first investigated in Alexis St. Martin, a man who received a musket wound in his Stella stomach. The hole remained open for fifty eight years! and his doctor was able to dangle little pieces of food on string inside and see what happened to it and also analyse the juices produced. Beef was digested quicker than chicken!

Stella stomach can also contract suddenly if she detects you have eaten something bad or too much which might even kill you, so you are sick and back out it comes. There are some disorders where people can be sick up to twenty times a day being triggered by various phenomena. Romans at grand feasts used to tickle their throats with a feather so as to vomit so that they could continue eating. Horses cannot vomit, nor can rats which is why rat poison is so effective. When humans vomit they dribble first to protect their teeth from Stella stomach acid. Birds can regurgitate indigestible materials. Stella stomachs lining

changes every three days usually. Cells are replaced quickly.
Thirty to forty thousand cells are lost per minute. She can
replace five thousand cells per minute. But did you know
that when you blush so does your Stella stomach-not very
obvious though unless there's a hole on your Charlie
chest?

Not much can be absorbed through Stella stomach into
the blood. Alcohol and water can but only if it has the
same salt concentration as in blood. Water can suppress
appetite and also help later on to metabolise the fat eaten.
If food stays in Stella stomach it becomes a soup. Sugars
spend the shortest time inside followed by proteins then
fats which float. At the lower end of Stella stomach a
pyloric sphincter valve opens and squirts the soup into the
first part of Gillie gut, the small Iris intestine (6.7 metres
long and just over 5 centimetres diameter - she's got a lot
of Gillie guts that Stella). This soup from Stella stomach
soup spends about one-two hours in Iris intestine.

Peter pancreas, a soft leaf shaped organ and only eighteen
centimetres long has tubes connected to Gillie gut, Peter
lends a hand in digestion by throwing in one and a half
litres of fluid which includes a cocktail of enzymes
including the protein digesting ones. Louis Lipase enzyme
has a go at breaking down fats. Carly carbohydrase
enzymes chop up the sugars and Privy protease breakdown
proteins. Juices from Peter pancreas also shove out into
the Iris intestine an alkaline juice (water plus bicarbonate)
to neutralise Stella stomach acid so that her enzymes can
work. So if too acid you've had it chum no food chopping
today the enzymes are on strike. Some plants e.g.
nepenthes also produce enzymes which eat birds and
rats which fall into their pitcher plant bowls. Digestion in
humans increases and doubles the rate for every ten degree
rise in temperature which means enzymes are working

better. In pythons the enzymes can dissolve Billy bone, Slinky skin etc.

Lily liver weighs one and a half kilograms in men. Lily has four lobes and removes debris and dead cells. She has cells arranged in a spiral around a central vein. You cannot do without Lily liver who works as a waste incinerator, a metabolic magician. It stores, recycles and releases as well as processing nutrients at a steady rate. Break down is increased with exercise. She can perform over 500 functions. As well as being a major player in chemical reactions and really full of red blood cells like myself Lily liver cells lend a hand throwing in bile into the gut via Brenda bile duct which runs from Gordon gall bladder which contracts.

Gordon holds up to sixty millilitres of the green liquid, Barbara bile. Three hundred milligrams per day originate from my dead red blood cell friends, together with salts, bicarbonate and cholesterol. Barbara's job is to act like washing up liquid so as to break up greasy fats and oils into little balls in the small Iris intestine so that the enzymes can run in and attack them better. We're having a ball said the enzymes when the bile arrived? If sick on an empty Stella stomach or when pregnant, you can often bring up bile. Women tend to burn off fat slower than men.

Lily liver is replaced every 6 months and you can remove eighty percent of Lily. She will regenerate. However if Lily liver is really damaged (cirrhosis) then scarring is permanent and no healing occurs. Drinking less caffeine and alcohol means less work for your Lily liver and less damage. Lily liver in the past was taken out from sacrificed animals (hepatoscopy) as well as other organs, to solve national and local problems depending on her shape and colour.

Iris intestines, if pulled out and straightened would be 5 times your height. This was often on display when victims were hung drawn and quartered. The drawn being Iris intestines pulled out. A Samurai warrior if defeated would commit hara-kiri where he would cut open his abdomen with his sword and pulled out his intestines and even cut of his own head! Peristalsis muscular movements helps as in other parts of the digestive system to push solids and liquids along Iris intestines Food can move at five-eight centimetres per second. The small Iris intestine is about 6.7 metres long but if pulled out can reach 300 metres in the body. Iris intestine also produces the enzymes for the finale of food breakdown. The ultimate chopping down is breaking bonds to form simple sugars, amino acids and fatty acids and glycerol. These really small molecules are left and can be absorbed by the later sections of the small Iris intestine .

The duodenum section is twenty five centimetres long. The next sections the jejunum and ileum are six metres long. Other materials are taken in such as alcohol, minerals and vitamins. Vitamin D is needed for calcium absorption and vitamin A is for pigment in the colour cones in your Ellie eye. In babies after birth and for some animals including fish for the first month large proteins can be taken through Gillie gut into the blood from breast milk. This provides the baby with blood proteins for immunity. This absorption only lasts for a few months. Break down of complex sugars to glucose helps Brian brain survive as this is the only energy source Brian brain can use. Do you want sugar in your tea? Yes please my Brian brain needs it for sweet dreams.

As a red blood cell I am travelling onwards towards Smartie small intestine. I will be inside the very fine finger - like projections of Victor villi. Victor is one millimetre

tall and on his finger like sides has around six thousand
little projections. If you spread out the insides of Iris
intestine with Victor villi and microvilli included then these
would be two hundred times the area of Slinky skin. The
total surface area of the entire Iris intestine if flattened out
is a massive two hundred square metres. As a red blood
cell while absorption takes place, I may be bombarded by
nutrients entering my plasma. Fats though are picked up
by other vessels in Victor villi lying next to me called
lacteals and go to the Lesley lymphatic system. Eating
more calcium as in yogurts can cling to fat inside your Iris
intestines and form soap so fat cannot be absorbed and
more fat is excreted Also there are cells all along Iris
intestine which secrete alkaline juice to reduce acidity so
enzymes can work. Some reptiles and snakes can takes
ages digesting food. Turtles may take 3 weeks. The big cats
often take infrequent large meals so make the most of
them. Lions in the wild only have about twenty kills per
year, perhaps dinosaurs did the same?

Primitive man also made the most of food found. .This
helped to build up reserves. Humans probably gained fat
in summer and lost this by eating less in winter. It's said
the presence of mammoths helped your survival by
providing you with food and clothing to keep us warm
during the ice ages. The male emperor penguin has to
survive for four months without food while incubating its
Elsie eggs on its feet in sub-zero temperatures. If it drops
Elsie egg the developing chick will freeze solid and not
recover. The Iris intestines of corals can exude and eat any
other coral that is a rival.

Some foods are toxic to humans but tolerated in small
amounts and over long term the body can be resistant.
Some animals unlike you, have enzymes to break down
foods which are poisonous to humans. Raw beans for
humans are toxic as you don't have enzymes to deal with

them as in uncooked kidney beans. Beans cause farts because of sugar raffinose which you cannot use. This is also in sprouts and cabbage – a good excuse for not eating them for some children?

Smartie small intestine then leads into Leslie large intestine. Leslie is 1.5metres long and shorter than the small Iris intestine but wider and with two bends in it). At its junctions there is April appendix. In humans April appendix is very small like a worm about ten centimetres in length but can be as long as twenty six centimetres. It's not much use to you now as you don't have lots of Bronwyn bacteria inside it to digest cellulose in green food like rabbits etc. It contains lymph tissue and may be antibacterial.

Leslie large intestine is where water, salts and potassium are sucked up back into the blood, so here I could get a salty bath. Animals that eat plants do not get enough salt so need salt licks, examples being elephants and even parrots. The body contains three buckets of water so this uptake of water helps you to keep this level. Normally you can lose up to 0.2 litres in the faeces. You need to lose more than 0.5 percent water before you start feeling thirsty.

If a person has diarrhoea (Leslies lining is not taking up water) I don't get so much water into my surrounding plasma as this all flows out of your Andy anus! Healthy people also have Bronwyn bacteria in their Leslie large intestine which produce vitamin K and B which you need. Babies, when born, don't have these Bronwyn bacteria but pick them up later. At birth babies have an injection of potassium in the foot, so that's why they cry. Possibly way back in the past you ate your mum's faeces to get Bronwyn bacteria. This is how baby rabbits, hares and rats eat their parent's faeces to get the cellulose digesting Bronwyn bacteria, which they keep in their April appendix.

Food and chemicals that enter next to me in the plasma in the blood system in the small Iris intestines are taken to cells and Lily liver where they are processed. Lily liver, as well as storing up excess glucose in a meal also deals with fats and amino acids, breaks them down and modifies them. Lily also breaks down toxins, poisons, Bronwyn bacteria and provides heat. It also stores added vitamins. Any food left in the Leslie large intestine travels to be stored in Rosie rectum where it is released through a valve at Andy anus as poo. Rosie rectum has two circular Mollie muscles. The inner muscle is non-voluntary while the external one is voluntary. When Rosie rectum is full the Mollie muscle relaxes and the external sphincter contracts more urging you to go to the toilet. If you cannot go for some reason then Rosie rectum expands. When its contents reaches two litres capacity then its poo out you go. (Run). The temperature in your Rosie rectum is higher than under your tongue. So that's why you get steaming cowpats.

Your poo is full of old Gillie gut lining; plant cells walls (cellulose and fibre e.g. sweet corn and carrots), and lots and lots of dead Bronwyn bacteria. A third of what you eat becomes poo. It is estimated you can produce two to five hundred grams of poo per day, 1-2 million worldwide. You can lose as much as fifty kilograms solid body waste down the toilet every year. Babies produce their own weight in poo in sixty hours, a lot of nappy changing! Cows can eat over a quarter of a ton of green food in a day and lose four times their weight in poo. Elephants lose 22.7 kilograms of dung a day. Good job elephants and cows don't have nappies? Panda's poo forty times a day possibly meaning they need to eat a lot of bamboo in order to get enough nutrients.

The longest human poo recorded is about 3.7 metres long and 10 centimetres wide. Dinosaur poo could be 62 centimetres long. The poo of reptiles can repel other reptiles and it is believed it was the same for dinosaurs. If you poo on the ground you can get a snake like coil of poo. Wombats have poo shaped like sugar cubes. Buzz Aldrin was the first to poo on the moon. Here poo is freeze dried and stored so it is not floating around above your head.

A poos appearance- colour etc. can be read against the Bristol Stool chart and can indicate health problems. A value of six is a fluffy with ragged edge, mushy looking poo. Vegan's poo is often very smelly. Baby's first poo or meconium is green, black and sterile. The brown colour of your faeces is due to red from the breakdown of my haemoglobin (bilirubin) and yellow and green from bile.

 Bronwyn bacterial breakdown changes bilirubin to darker urobilinogen. Around the end of the 12th century there still were no toilets so men dressed in very large capes with a bucket in their hand would wander around the streets. They spread their capes, around customers who sat on the buckets to poo or wee. Accumulation of poo in crowded conditions can lead to disease and attract vermin. In castles, small boys had to clean the poo from the long vertical shafts which ran down from the top to the river.

Through your anus you can produce a pint of gas per day. This is made up of hydrogen, methane nitrogen, carbon dioxide and hydrogen sulphide if you've eaten meat or Elsie eggs (eggy fart). Humans can produce six hundred millilitres of gas per day. Adolf Hitler ate vegetables and had flatulence which caused cramps. His doctor gave him capsules containing good bacteria from the faeces of German soldiers. Poo can be smelly especially when on non-meat diets. Vegetarians can fart more. Some foods cause a characteristic smell at both ends e.g. onions, garlic or curries. You can even rub garlic on your feet and you can smell it in your breath.

You can fart fourteen to twenty times a day. With Crohns disease you can fart even more. Seventy farts have been produced in four hours. Usually women fart less but they are more stinkier. Cockroaches can fart every fifteen to twenty minutes. Even herrings are said to fart in the sea. Gillie gut produces many explosive gases, so take care. Competitions exist for the furthest flaming farts. In sheep over eating of clover plants may cause Stella stomach to blow up like a balloon. You have to puncture Stella stomach to let the gas out. If you like music humans can also play music or musical instruments using farts. They can even blow out candles.

Cows produce five hundred litres of Gillie gut gas per day, not by farting but by belching. Orang-utans tell people to keep off their territory by burping.

In the past, as mentioned previously, it is believed that your ancestors ate meals only at intervals. It has now been shown that if you still did this now and did not continually eat all sorts of foods every day then your bodies would have time to get rid of toxins so that there would be fewer cancers etc. In the past you tested everything, many people dying or suffering after trying them out. The Japanese Puffa fish is an example which has a paralysing poison in its Lily liver which can kill many people.

Old people were used to prepare new foods and eat them to test if they were poisonous as they are more expendable compared to the younger people. Some royals and even Hitler had tasters to check if food was poisoned. Not the best job in the world. You eat all sorts of things even ourselves, even sold as meat in sausages and pies. Cannibalism was common in the past and many enemies were eaten. Sailors used to eat meat in casks which was over 10 years old. Meat in some countries is often preserved by drying. In some countries you can eat spider salads, rats/bats and guinea pigs not to mention thousands

of insects, worms and grubs. Cockroaches in turn can eat
your sweat and Slinky skin in any human orifice. Insects
can be made into soups and even eaten in lollipops.
Millions of flies around Lake Victoria in Africa are made
into burgers and have a high protein content. You ate
many different birds and other animals especially during
war time. People have survived in starvation conditions on
eating grass and will eat shoe leather. The first recipe
books for safe foods were produced or handed down by
Moby mouth.
Even now different people/ tribes eat different things.
Examples include fish and other animal Ellie eye balls and
live monkey Brian brains. Wriggling octopus tentacles are
especially popular where suckers remain functional and
stick to the inside of your Moby mouth. Other delicacies
include Tony testicles (annual testicle eating competitions
exist), animal penises, the nests of swifts and milk with
blood drunk by the Masai in Kenya There are
competitions for eating certain numbers of insects.

Some people eat innominate objects such as ash, glass,
metal, and their own Harry hair. Light bulbs and Cessna
plane and bicycles have also been eaten (in bits that is, as
Moby mouths are not big enough!). Let's hope their acid
Stella stomach is top notch, and the hydrogen they
produce well away from a naked flame. Some people
swallow swords which can damage Gillie gut. Fifty swords
have been recorded at one time. Prospective sword
swallowers start by putting their fingers down their throats
and then sticks to overcome the gag reflex. Sometimes
objects and food can lodge in Stella stomach. Vomiting is a
safety feature of the body. With an attack by Ebola
Veronica virus your sick is black. Vomiting is not caused
by Stella stomach but caused by voluntary Mollie muscles
of Dickie diaphragm moving suddenly upwards with the
abdominal wall rigid. Some disorders can cause people to
vomit over 20 times a day triggered by different events.

This varied diet caused lots of different chemicals to surround me but your Gillie guts and body systems can adapt to anything thrown at them within reason. Some foods release different amounts of minerals e.g. vegetables have less iron than meat.

Not all hunger is real hunger; it may be you need water. So drinking water can stop you feeling hungry sometimes. Eating breakfast helps to kick start your metabolic processes for the day. Eating large carbohydrate molecules e.g. starches in cereals or porridge, means that it takes longer to break these down so that you have extended energy supplies. Highest protein in a fruit is in an avocado. Just eating or drinking something sugary, especially if it's glucose, gives you a sugar high (instant energy boost) but you soon come down with a bang and can feel tired and depressed. You need to eat more fibre and protein at meals and lots of water. Too many calories can be redirected to fat.

The basal metabolic rate (BMR) for women measured indirectly by carbon dioxide out and oxygen taken in is six thousand three hundred kilojoules per day while for men it is seven thousand one hundred kilojoules per day. Half of the energy from food is used (metabolised) for general body functions. At rest as much as three hundred and sixty kilojoules can be generated as heat. The rate of food breakdown depends on your activity. Even if you are not active Happy heart and Brian brain etc. are still working. This is called the basal metabolic rate which is measured indirectly by how much oxygen you need to take in and how much carbon dioxide you breathe out. It can be measured when you are lying down at rest, have fasted (not eaten for twelve hours) and your body temperature is normal.

Gut problems

Several problems can be found in Moby mouth. Common Moby mouth problems can include sores and gingivitis

Bronwyn bacteria. Sometimes unusual things can be
swallowed and some creatures have been said to live inside
your Stella stomach.

Termites and even 30 centimetres long water snakes have
been mentioned. How true this is, is questionable. Worms
are common in Iris intestines Fifty-six worms were found
in a Japanese person's Stella stomach. Some worms can
produce a million Elsie eggs per day.

Stella stomach ulcers can be due to acid damage to the
lining; this then gets infected by Bronwyn bacteria
producing heartburn and a dull ache a few hours after a
meal.

For some disorders sections of Stella stomach can be
removed as is seen for Iris intestines, but the patient would
need drugs to slow down the passage of food and
absorption. Gastric bands partitioning off parts of Stella
stomach have been used. These can be adjusted from
outside and have been used to help lose weight.

Gall stones can be a problem blocking movement of
Brenda bile from the bile duct or functions in Lily liver.
They can be various shapes including stars. If stones enter
Lily liver it blocks the bile duct so overflows into blood.
Bile is yellow and darkens urine. Gall stones, if large can
reach six kilograms and in some cases can have more than
twenty five thousand small ones.

Intestinal problems include Crohns, constipation,
diarrhoea, irritable bowel, sphincters not working and
gastric or duodenal ulcers in the lining of Stella stomach In
Crohns disease the lining of the small Iris intestine is
ulcerated and damaged. This reduces Victor villi so fewer
nutrients are absorbed. Also less blood and lymph goes to
Victor villi so there is less iron uptake. Where serious
constipation can occur you can get reverse peristalsis
where faeces can escape through your Moby mouth.
Irritable Boise bowel is increasing often occurring in early
adult life, especially in women. It results from over activity

89

of Gillie gut brought about by stress, food sensitivity but not an allergy or possibly an infection triggers it. In coeliac disease the person may have intolerance to gluten, the protein in wheat. If they eat this the immune system attacks the small Iris intestine so that no materials can be absorbed so one can get weight loss and in extreme cases diarrhoea and vomiting. as a body defence mechanism.

Lily liver and Peter pancreatic disorders can affect digestion also Cirrhosis can be. Ground up Egyptian mummies said to make you stronger but not easy to get hold of these days caused by alcohol or hepatitis. Lily liver may be scarred after damage
Lily liver cells are replaced every one hundred and fifty days. They do not age and if you lose half of Lily liver it can regenerate in three months. You can also transplant an older persons Lily liver to a younger person. If Lily liver is not working this could be due to a Veronica virus attacking the cells e.g. hepatitis, Lily liver damage is indicated by jaundice which results from a break down product of iron pigment in the blood (bilirubin) This deposits in Slinky skin and Ellie eyes as yellow deposits. Hernias can occur in any part of the Iris intestine including Stella stomach. These are where portions of Gillie gut protrude through Mollie muscle walls and food transport is affected and Bronwyn bacteria can thrive in these pockets.
Haemorrhoids can be found in Andy anus and Rosie rectum. They are vascular structures which cushion the anus and can become enlarged. These can bleed and are sore when passing faeces.

Some animals and insects or parts of, if eaten whole or ground up into flour, were believed and still are in some cases good for ailments. Examples include foxes for Lucky lungs bats for vision Doctor Thomas Mouffet gave his daughter spiders on toast as a remedy for intestinal problems. This is how the little Miss Muffet nursery rhyme came about. Ground up Egyptian mummies are said to make you stronger but are not easy to get hold of these days. Overeating can kill you and some kings have been killed by this.

Cancers

Cancer cells are immortal, never dying, and given the right conditions would continue to grow and divide for ever and ever. This would be great for you but by dividing they get in the way and block body systems and cause chemical imbalance. They are commonly found in similar dividing cells such as Gillie gut and Slinky skin.
There are hundreds of causes of cancer. Some include viruses, alcohol, radiation- X-rays/gamma rays, diet, tobacco, sunlight UV, occupation, chemicals and hormones.
Smoking followed by alcohol is a common cause of cancer not only to Lucky lungs but Moby mouth and throat and the tongue. Tongues may have to be cut out and may be replaced from Mollie muscle from the inner arm, wrist or throat. Lymph Gladys glands which may transport the cancer around the body are also removed. Cells in Slinky skin in the Iris intestine are rapidly dividing and can be affected easily and carry the defect into the new cells.

Eating more fruit especially blueberries and vegetables, with their antioxidant properties prevents oxidants damaging the cell membrane and Diana DNA. Some wines and beers also contain antioxidants which are believed to help cells repair themselves better if their Diana DNA is damaged. Cancer caused from radiation by exposure or eating contaminated material can kill off your Winston white cells. In Hiroshima when the bomb was dropped many people exposed to the radiation lost their Winston white cells leaving the body defenceless. Parts of the body began to become infected and die (necrosis) seen as brown spots, on Slinky skin.

More serious cancers are those of the Lucky lung, Owen oesophagus and Peter pancreas. Cancer growths are removed surgically by lumpectomy where surrounding tissue and lymph nodes are removed. In a mastectomy all breast and lymph nodes are removed. Cancer cells can spread into the blood and Lesley lymphatic system and lymph nodes and cause small growths in any part of the body which will cause death. Billy bone tumours where blood cells are made are especially a problem.

Lasers to burn out cancer cells or freezing cells are also treatments as well as chemicals and enhancement of your immune systems. Cancer cells, as they are growing, need food and oxygen so if you can restrict the new Christine capillary systems that may be produced to feed the cancer you may be able to starve it. Parts of Gillie gut can be removed and successfully joined to other parts. Stella stomach itself can be removed and the person can take drugs to help with digestion and retention of food in the Iris intestine.

Treatment with radiation especially with a high dose may kill off non-infected surrounding tissues so it is done in several stages. Radioactive material can also be placed inside next to the cancer. If a cancer is not seen after 5 years then it is likely it may have gone. A more target form of radiation involves proton therapy mainly killing cancer cells. This is often used for brain tumours. Introducing Jenny genes into cells is known to shrink some cancers. Ultrasound and lasers are also commonly used for cancer removal.

Alcohol

Increased use of drugs can also increase the chances of cancer and kidney disease.

Alcohol can expand blood vessels and reduces blood flow. Twenty percent of the alcohol taken is absorbed through Stella stomach, the rest in Gillie gut. In an empty Stella stomach alcohol is absorbed quicker. Drinking milk slows absorption. Alcohol kills cells, and enzymes don't work.

Alcohol is slowly metabolised by Lily liver, and can eventually give energy. This is faster if a person has not eaten. Too much alcohol can eventually produce scar tissue, cirrhosis, and this cannot rejuvenate. Lily liver cells can be created from stem cells using chemicals and hormones but only can be used if Lily liver is not too far gone. Note it is better to drink alcohol 4pm-11pm (happy hour?) as it is broken down better then. Alcohol also affects motor co-ordination (movement), the size of ventricles in Brian brain (an increase), reduction of vitamins B6 and folic acid especially. Other effects include delay in messages to Brian brain, ulcers, nausea, and smaller babies. and irritation and inflammation of Iris intestines.

Dilation of blood vessels caused by alcohol lowers blood pressure. In slinky skin in the face this can cause reddening and overheating. Pigs can also be alcoholics and sheep can get drunk and stagger around

Drinking also means you don't go into a deeper sleep phase. Also if out late at night the longer it takes to cool down. Babies born to mothers who drink often have flattened more rounded faces and other features such as Noddy nose is less prominent Also men that drink have lower Speedo sperm counts and often children born tend to be girls.

However 1 or 2 units of red wine or even tea can protect Happy heart. Also some spirits taken by older people have the same effect. However over 10 litres per hour can kill you or nine to ten drinks in one hour. Alcohol kills and preserves cell and is used for pickling. Even large animals are preserved in alcohol in museums in what is called the spirit room. Nelsons body was preserved in a barrel of rum which was drunk by the sailors. In Canada a human toe was included in a drink which you had to taste before becoming a member of a society. The alcohol is believed to kill of any bugs from the toe just as drinking communion wine can kill off germs passed on by those drinking from the same cup.

Eating disorders

Eating less can help you it is believed live longer but not fasting to extremes. Two of the major eating disorders are anorexia and bulimia. Often the second follows on from the first. With modern day pressures on appearance, especially for the younger generation, both of these disorders are more common in women and can run in families the children being influenced by the mother. Both are a fear of fatness.

Anorexia is under eating and those affected can also be overactive or over exercise on top of this. They may also eat mainly a lot of fruit and vegetables. Anorexia nervosa in extreme is not eating very much at all whereas bulimia is eating but making ones self-sick. In bulimia they can eat a lot of fattening food then unknown to others vomit this up. This has the problem of bringing up very strong Stella stomachs acid into Moby mouth so that their tooth enamel is affected and teeth eventually dissolve. They also can have swollen Sally salivary glands. Fasting can reduce the amount of enzymes produced in your Gillie gut by ten to fifteen percent of normal. The smallest waist ever recorded

is 38 centimetres. There are drugs which block the
absorption of fat but you need to take care if slimming.

In women both disorders have a large effect on their
general health, affecting the strength of Mollie muscles,
Billy bones and their reproductive capacity as well as some
cases leading to death. With the reduction in Mollie muscle
mass and fat they also feel the cold. In the past often
people starved themselves to make money Claude Seurat
(1798) only ate a roll and a glass of wine every day and you
could see his Billy bones as a skeleton and his Happy heart
beat was pronounced. Even so he remained healthy. There
were only seventy six millimetres from his Charlie chest to
spine and he had to be careful not to be blown away.
There is an MDP syndrome also that exists where the
person is mostly Slinky skin and Billy bone with little fat in
their body. The human skeleton common in side shows in
the 1930's, weighed just twenty six kg. As with tradition he
was allowed to marry a fat lady so as to increase public
attraction and publicity.

On the other extreme you have starvation in poorer
countries or the slums of some of the richer countries.
Kwashiorkor is starvation where children do not have
enough protein so that fluids leak out into the blood and
tissues, especially in the abdominal cavity which swells up.
Marasmus is a similar disorder, but this is where children
are not suddenly experiencing starvation. Here they have
been deprived of sufficient food from birth.
In starvation conditions in the past it is known that some
races ate women before dogs as dogs were more useful.
Geophagism where dust or clay is eaten, is common
around the world. Many Orientals and black Africans do
not have an enzyme to break down sugar in milk. i.e. lack
the enzyme for lactose breakdown. They can though take a
supplementary enzyme. There is a syndrome called Prader
Willi syndrome where the person can never stop eating.

Additives
Food additives such as flavours or colourings often include chemical dyes which are linked to cancers. These are abundant in soft drinks, sweets, lollies and cakes. The problem is they are very attractive and may mimic the true natural colour of a food or fruit which encourages parents to buy them for children. One example is the use of colouring and dyes in birthday cakes often iced to represent a latest children's film or story. Colouring can also alter the behaviour of some children making them hyperactive. These products have an effect on the development of Brian brain up to seven years plus. Vitamins in excess are toxic and many meats contain anabolic steroids which can be absorbed by butchers through Slinky skin and affect their weight. Allergic reactions can be caused by allergens which can be chemicals, proteins etc. These can be in foods, pollen, stings, even dust mites. They and can cause an immunological response resulting in the overproduction of histamine from mast cells in connective tissue and white blood cells. Blood vessels can dilate so increasing the response. Anaphylactic shock can result. Minute amounts are needed to trigger a response and an example can be allergies to nuts where a partner has been eating nuts and kisses someone allergic to them.

Constant exposure is a problem and may cause sensitivity to many chemicals. Latex can cause allergies as can shellfish and even human Speedo sperm. Toxic body necrolysis is a total body allergic reaction where the top layer of Slinky skin sloughs off on sheets and leaves the body exposed to infections. With Victorian match girls, phosphorus caused phossy jaw. Here the gums rotted and teeth fell out and phosphorus destroyed Billy bone. One

could always tell a match girl as when she was sick in the
gutter the sick was luminous.

.

Famine and Obesity

There are more overweight than hungry people in the
world. Every day over 30,000 children in Africa die from
starvation. The country most affected by famine is Sudan.
Forty six years of age is the highest life expectancy in
Africa while in the western world it can go beyond one
hundred years. Pear shaped women tend to have less heart
disease lower blood pressure and less diabetes than apple
shaped women.

Fat is stored in young animals throughout Mollie muscles
but mainly on the outside older ones. For every 10 grams
of fat in a meal a woman can store 4 grams in their
subcutaneous tissue while men only store two grams.
Eating high fat and high sugar meals can make you sleep
on long journeys so can pose a danger. The more fat that
there is in the body then the less water present. Some
people can weigh over five hundred and sixty kilograms
and need a lot of people to move them. In some cases
their hips can reach many metres. They can be bed ridden
for many years and may not be able to get through the
doors and may need the windows taken out to get them
out of the house. Mollie muscles can weaken and bedsores
can become infected and lead to sceptic shock. Weakening
of Charlie chest muscles can lead to trouble with breathing
and with fluid accumulation in Lucky lungs with
development of pneumonia. Extreme prolonged sitting
may cause their Slinky skin to be attached to a chair.

The presence of fat increases buoyancy in water so fat
people float better. Women have more fat than men so
survive longer after a ship wreck. This buoyancy is
illustrated in bottom dwelling fish such as skates that have
less fat in their Lily livers so that they sink. Some whales

have over fifty thousand kilogrammes of blubber which helps support them. People who are very obese can have problems with food passage through Gillie gut and food may be trapped and rot. Ultrasound can break up fat this can be removed by liposuction.

Obesity has tripled in the last twenty years and the consequences of this for future generations are worrying. Sumo wrestlers have subcutaneous fat (just under Slinky skin) and not visceral fat (around organs) so it can be used up quickly. They can eat over thirty thousand calories per day to increase their weight. Fat can be quickly used up if they stop fighting at an early age and become thinner. Obesity can make you less fertile, and gave you a tendency to diabetes, gallstones, arthritis and heart disease. Lily liver of fat people can be engorged with fat and be grey or pink. Other organs may adapt to high fat levels. You produce a lot of mucus to get food out fast.

Some people/races are naturally overweight or obese especially in some of the warmer countries. This may be considered a sign of beauty or affluence (rich) Some tribes have women with big buttocks which is very attractive to the opposite sex. Examples are Hottentots. Hottentot Venus was found in side shows and displayed in the Egyptian Hall in Piccadilly Circus in London. A thin European seen by these tribes would be considered to be really ill!

Some husbands in some countries prefer their wives to be big and parents will force feed their girls from a very young age with milk and starchy foods, even restraining them so that they don't vomit. This feeding can cause rapid growth. Care has to be taken not to eat meat and milk as the calcium in the milk is not absorbed as fast. Some women can reach a natural breast size needing a

102ZZ, weighing 44.45 kg. Implants of course can increase size massively.

The arrival of food chains selling junk food is a greater problem for these people who already are overweight. Calories required for maintaining a healthy body weight vary a lot depending on your physiology, build and lifestyle. Whilst sleeping you can use fifty five kilocalories per hour, while running use four hundred kilocalories per hour.

It is your children who are suffering with fast foods and it's not only the lower class but upper and middle are affected. High sugar, fat and salt levels are the problem with lower levels of vitamins and minerals. It's a pity you don't have tails where you could store your excess fat like some of the Australian mammals.

If your body mass index (BMI) weight in kilograms divided by height squared in metres is over twenty five then you need to lose weight. If its 17.5 kilograms per metre squared then you are anorexic and need to be careful.

Weight measurement is believed to be more accurate as BMI does not take into account Billy bone structure and build. A waist measurement of greater than 81 centimetres in women and ninety four centimetres in men can mean you are at risk. If you get very large e.g. over two hundred and three centimetres waist then Bronwyn bacteria which like fat, can get into sores produced by friction and it becomes less bearable to cope.

Very overweight people in the past were more of a rarity and were put in circuses, fairs or drawn in pictures as in a stupor or asleep. Now they are fully accepted in society. Some have tried to lose weight by using gastric bands which can be tightened. Parts of Stella stomach can be removed. The respiratory quotient is the volume of carbon dioxide produced divided by oxygen used. This is high where excess sugar is present as the sugar converts to fat releasing oxygen so less oxygen is needed to breathe in. Stapling Stella stomach is one extreme form of treatment. As well as numerous kinds of diets, drinking water can reduce appetite as well as the nice smell of food can help stop you eating. However too much water makes your cells swell and you can get Brian brain seizures.

Ageing and the Digestive system

Ageing reduces the effectiveness of organs and secretions (enzymes/ hormones). Therefore you need a healthier diet and exercise to maintain fitness. After fifty five in the Leslie large intestine you lose lots of friendly Bronwyn bacteria. These can be supplemented with bioactive yoghurts. Vitamin supplements can also help enhance reduced absorption in the gut which occurs with age. Free radicals in your food and in the air have been bombarding your cells and their membranes all your life. The free radicals are unstable molecules of oxygen which damage Diana DNA so that it produces more errors. At seventy five you could have lost sixty four percent of your taste buds that is why older people often tend to prefer spicy, more flavoured food perhaps.

Also at seventy five-to eighty years old your Gillie gut starts to shrink. Metabolism is affected by the foods you eat. Meat requires a lot of energy for breakdown. The higher the metabolism the more ageing that can occur. In some countries where they have a small basic diet e.g. rice, vegetables and fish or meat now and again, people tend to live longer. Salt is essential in any diet, without it you can die in three weeks.

Irritable bowel is common in the elderly. In some cases part of the Iris intestine can be removed. The toxic effects of some foods are reduced by cooking Cassava is an example. Older people in some tribes deal with cassava

which contains cyanide, but cutting can be dangerous. Old
people do this as they have had many years whereas
younger people are just starting their lives.
In the elderly with problems with constipation etc. food
can remain in the Iris intestine and rot which causes
problems.
Some elderly people especially in the past spontaneously
combusted (i.e. burn up leaving only Billy bones This
often depended on flammable chemicals in the blood
either produced by illness or by eating ,or drinking alcohol.
Sometimes only small parts of the body ignite.

8 CLEAN UP JOB FOR KATY KIDNEY AND NICKY NEPHRON GETS A SIEVING JOB

Some animals have no urinary system e.g. sharks leak urine through their Slinky skin. At last in the plasma with Katy Kidney's help around me I have some relief from all the waste that was travelling within me. Your body is 70-75 percent water and Katy kidney maintains water and salt levels whatever is eaten or drunk.

Katy Kidneys are a pair of dark red lobes (eleven centimetres long 6 centimetres wide) and can process one hundred and ninety litres of blood per day or 1.3 litres of blood per minute. Katy kidney with Lily liver are the perfect detoxifying machines and you do not need detox diets if you are healthy. There are over one million filters inside the kidney and all blood in the body is filtered every ten minutes and about one hundred and fifty times per day, a real achievement. It is therefore important that you drink water or fluids to maintain filtration as you will only survive a few days without them.

The blood is thicker so no more water is lost as sweat so you cannot cool down and you begin to have a fever and cook. You need to drink half a litre of water a day to excrete waste products but as water is lost in other ways you really need a litre. Drinking more water will make no difference .Compare a camel that can drink fifty seven litres of water without stopping. Water excretion with salt can also occur through Slinky skin by sweating. Cows as well as some other animals only have sweat Gladys glands

in Noddy nose. If you live in a very damp climate means you sweat a lot at first then sweating stops. However living in a very dry climate you also can adapt and lose less water but sweating still occurs at very high temperatures.
Drinking 2 litres of water a day can burn off one hundred and fifty calories

Slow, slow, quick, quick slow is all I can say about passing through Katy kidney.
Katy kidney is basically a collection of millions of small tubes called Nicky nephrons. Nicky nephron's family have a filter at one end to filter the blood. The majority of fluids in the blood pass through in a small knot of Christine capillaries and pass under very high pressure through the filter into the tubes. Large proteins and the blood cells do not pass through the filter as they must not be lost.
A lot of waste in the body is toxic, mainly from excess amino acids, chemicals and salts you eat from food. Fish can excrete ammonia which is toxic to you. This in fish will dissolve in the sea or rivers. Birds get rid of their toxic uric acid by dropping it from the sky (need to remove it as it burns cars not to mention the bugs). Birds also have Gladys salt glands in Ellie eyes or nostrils which enable them to excrete salt. The stormy petrel will shoot salt at you through its nostrils if you approach its nest. You get water from food but some animals rely entirely on their food. Koala bears do not drink but like some desert animals, they can extract water from eucalyptus leaves. Kangaroo rats get their water from metabolism from eating seeds etc. Some animals e.g. kangaroos have a lower metabolism than you have. This means that they do not need as much water to survive.
The initial part of the tube after the filter in Nicky Nephron reabsorbs nutrients which are useful, while further along drugs and old hormones are secreted into the tube The later parts of the tube (in a Larry Loop of Henle) are linked to exchanges of salt and water to enable the

levels to stay the same in the blood and cells throughout the body. Glucose is reabsorbed so there should be none in the urine

The hormone ADH (Antidiuretic hormone) allows more water to be taken back in from Nicky nephron's tubes if the blood is too thick, so a lot of water is not lost. Less of this hormone (which alcohol encourages) means that more water can be excreted so that you wee a lot. Receptors in Brian brain and in large Vernon veins monitor water content and detect less water in the blood. This regulates thirst. Adults need two litres of water a day. This is important, especially if you drink over forty five litres in a life time.

Nicky nephron's tubes lead into the cavity in the kidney and Elvis pelvis, and travel down a tube, Rita the ureter, which contracts and force urine down into the back of Betty bladder where the urine is stored before release. Betty bladder can hold six hundred millilitres of urine. At one hundred and fifty millilitres the Nellie nerves in the walls of Betty bladder tell the brain you need to wee. The male Betty bladder is higher in Elvis pelvis than the female. The urine exit tube, the urethra for is twenty centimetres in men and only 4 centimetres in females. Normally you can wee 1.4 litres of urine a day. In a life time Katy kidney's urine can fill five hundred baths or a small swimming pool. Kidney function and urination is reduced at night as your urination and sleep cycles are out of phase. You wee less between 0400-0600 am (you should be asleep anyway?) and wee more 1200-1400 pm (mid-day to early afternoon). Traces of alcohol reach the urine within 12 hours.

Eating too much salt is a big problem today in all foods and causes the body to retain more water making you more puffy and swollen. You excrete nontoxic uric acid and other breakdown products of nitrogen (urea and creatine from Mollie muscles). The urine has a lot of water in it to dilute the toxins in your blood. Urine is yellow due to excretion of bile. It is more yellow after sleep and some foods make it darker than others. Cat's urine glows in the dark and some small mammals urine contains chemicals that glow in UV light. Kestrels can, unlike you, see this and know if these animals are about.

Diogenes, a Greek scholar was well known for urinating on people he did not like. Howard Hughes, a billionaire, used to collect and label his urine and keep it in the fridge. Buzz Aldrin was the first man to pee his pants on the moon but I doubt if it ran down his leg. In space urine forms crystals and looks like sparklers. It is normally collected and bagged and freeze dried.

Drinking acid urine of a pregnant woman can increase your immune system it is said but I don't think everybody would fancy a pint of this in a pub! The urine of young girls is supposed to enhance the flavour when added to cheese. Urine also has healing properties. Many Hindus still believe cow's urine has this property. The Romans brushed their teeth with urine and gargled with pee. Urine has been collected and used in tanning and softening leather and used with dye for uniforms. Also in medieval times, they hung clothes over urinals so as to keep moths away. Eating certain foods can make urine smell. Some people do not have a particular gene which affects this. Gunpowder at one time contained stale urine where potassium nitrate was extracted from it.

Exchange of fluids here in Katy kidney or exchange of gases in Lucky lungs depends on concentration differences between two places as well as differences in pressure. This causes molecules to move from an area of a higher concentration to a lower concentration, this is called diffusion. Like the lady in a shop who has a quick squirt of a perfume the perfume spreads around the shop, until finally it gets so mixed up with the air molecules that you cannot smell it any more. The more concentrated the perfume (more expensive Channel No 19) then the further and faster the molecules move and the longer the smell lasts. The same goes for liquids moving across the thin walls of the small tubes in the kidney. If water moves from an area where there is a lot of water (very dilute) to one where there is little then this is called osmosis. Salts are especially important for making cells in your body work, especially Nellie nerve cells and Mollie muscles cells. Water often follows salt but if the salts move across walls then you need a method to force the salt back again to where it came from. This process is called active transport and you need energy (ATP) to do this.

Kidney problems

Losses of more than 0.5 percent of body water causes thirst. Lose 8 percent of body water then you become dizzy. You need 1.2-1.5 litres of water per day to reduce losses. Urination can be once every three days to three or more every day. Katy kidneys you have seen regulate water and salts. Hormones such as insulin and glucagon regulate sugar levels in the blood. Insulin converts glucose to glycogen in Lily liver and aids Mollie muscles in using sugar. Glucagon converts stored glycogen to glucose again if the body needs it. Sugar should not be in your urine. Most people do not realise that they have a problem with blood sugar levels and may have diabetes.

Diabetes type 1 is seen at ten to sixteen years of age where the immune system can destroy insulin producing cells, so blood sugar is kept hyper (too much glucose in blood) giving a sugar high and glucose in urine. Diabetes type 2 often occurs when over forty years old where there is not enough insulin for the body's needs. If there is high sugar in the blood this can cause kidney failure. You also attract Bronwyn bacterial and fungal infections in broken Slinky skin and around excretory orifices.

Kidney stones are common in all ages but are often found in older people. They can have   many wonderful shapes and sizes and are caused by high calcium in the diet. The largest can be over one and a half kilograms. Faulty Katy kidneys can cause dehydration in the body. If your Katy kidney becomes inflamed, your temperature rises and you may have abnormal pain. Ultrasound shatters stones or they can be removed. Even very young children may get stones.

 Nephritis, which affects the little filters in Katy kidney, causes the production of dark urine and water remains in the body so that tissues can swell. Toxins are often produced with urinary tract infection such as cystitis caused by Bronwyn bacteria and Felicity fungi. This can cause irritation and pain and in serious cases sterility. The Urethra in women is shorter than in men so they can get more infections.

Some children are born with one Katy kidney or can even have them fused together. Also some valves don't release urine in Betty bladder but this gets better later. It is also known that some people are born with a kidney half way up their back or even with four Katy kidneys. With serious problems causing Katy kidney failure e.g. an infection or haemorrhage a kidney transplant is needed and they often leave the old one in place. Often the donor may be related so as to get a better tissue match. A person may have to continue with kidney dialysis (a machine that removes the waste from the blood but leaves essential proteins and nutrients). There are now portable small dialysis machines which can be taken on holiday the machine and filter cartridge linked to a catheter implant.

When the body has a cold or fever, often more water is retained for sweating and cooling the body, so that the urine looks darker. Blood or protein in urine could indicate kidney or Betty bladder disease. Nitrates in urine may be an infection.

Various creatures may get in to the urinary system. In India a boy produced a flying beetle in his urine and marine parasitic fish are known to enter the urinary exit Drinking over three litres of water in two hours may cause kidney failure as the kidney is overworked and less salt reaches the cells.

Ageing and the excretory system problems

Katy kidneys can shrink as you lose the little filters (Nicky nephrons).

Valves are not so efficient and old women tend to leak whereas old men block up due to Patrick prostate enlarging and pressing on Betty bladder causing frequent urination and pain in the back and hips. At around sixty five you can lose Betty bladder control. The Patrick prostate, originally the size of a walnut attached to the base of Betty bladder, can also become cancerous.

Gout is caused by uric acid crystals accumulating in joints (too much uric acid in blood). This produces stiffness and pain commonly in the big toe or thumb and legs. Birds especially those that eat meat can get gout in their legs and wings so cannot walk or fly. Some women have a fish odour syndrome (trimethyl aminuna) in sweat, urine and breath.

9 GLADYS GLAND AND RELATIONS TAKE CONTROL OF THE SITUATION

There are many Gladys glands in the body important for secretions of chemicals. These make things happen by affecting other Gladys glands or acting on body tissue itself. These chemicals are carried around in the plasma that surrounds me.

Hormones from the glands travelling in the blood help with the regulation of growth, repair, reproduction and the use of nutrients. Christine capillaries pick up hormones from neurosecretory cells (secreting Nellie nerve cells) from Brian brain. Small amounts of hormone produce a cascade effect and cause large changes as they hit their target cells. Targets may be Nellie nerves or other Gladys glands which have receptors, respond to hormones and these can produce enzymes, impulses or other hormones as a response. Each hormone structure and the receptors on the tissue match so that areas are targeted.

Hormones are broken down by Lily liver or the target tissue itself.

Harry hypothalamus is the main factory for hormone production in the body. Finger nail size it orchestrates nervous and endocrine (hormone) systems via Pippa pituitary. Pippa is a pea sized Gladys gland at the base of Brian brain which has lots of hormones. They all are released into my blood plasma in minute amounts but their reaction is slower than for Nellie nerves.

Pippa pituitary is pea sized and found in Brian brain. Hormones from Pippa pituitary also include FSH (follicle stimulating hormone) and LH (Lutenizing hormone) which help female follicles to develop. Pippa pituitary is also linked to growth and if affected one may get gigantism or dwarfism. ADH (antidiuretic hormone) is also from Pippa pituitary and helps with water regulation in Katy kidneys. Its presence allows more water to be taken back into the blood. Alcohol reduces the amount of ADH so more water is lost when drinking alcohol. That's why the loos are full in pubs. Nicotine (smoking) and morphine cause the reverse, more ADH less urination (peeing).

Pippa pituitary gland and Hazel hypothalamus are also responsible for producing endorphins which have the same effect as opiates in producing the feeling good factor. Endorphins give the falling in love effect and also help against pain in the body. During exercise endorphins are also produced. Beating someone's bottom with a stick can release endorphins and help reduce depression! Emotion and pain are found in the same region of Brian brain.

Other hormones in the team, but not in Brian brain include a well-known hormone, adrenalin. This is the one shot out from the triangle shaped Angela adrenal glands (5centimetres across) sitting on top of your Katy kidneys. This gets you ready for flight (action) and fight. It also affects metabolism and can change your mood from depression to elation and vice versa. For flight/fright it gets the body instantly ready by restricting blood flow in Slinky skin and Iris intestines so more blood goes to Mollie muscles so Happy heart beats faster. It also widens airways so more oxygen gets into your body. That is why adrenalin is given in anaphylactic shock or cardiac arrest cases. Adrenaline, when produced under extreme stress, can actually give a person sudden super strength so that they can produce heroic feats.

Other well-known hormones are the sex hormones. Testosterone is found in the male and produced from

Tony testes, while oestrogen in the female is produced from Olive ovary.

Testosterone peaks, it is said, at twenty or forty years of age. Maleness is decided at birth by the level of testosterone produced. Some women also have some testosterone which is responsible for maleness. Athletes, even women, have taken testosterone to boost their Mollie muscle development. Do you know any manly women? Men born without Tony testicles (sterile) are eunuchs. They are more woman shaped and have their characteristics but they still can get an erection. They are found in harems, not the erection but the eunuchs.

Young boys in the 16th and 17th centuries were castrated (Tony testes removal) in order to keep their voices high (soprano). This is because low levels of the male hormone would not allow the vocal chords to grow longer (Giraffes have no vocal chords! so they cannot sing!). These castrated boys were called castratos. The sound produced is not the same as a female's voice but is unique being between a child's and a woman. To be castrato could end poverty for you and your family. The child could go to a better school. Their voice lasted until old age and older castratos could act as mentors to younger ones. Even up to the beginning of the twentieth century castratos were still found in some churches.

Women and men, to a lesser extent produce the female hormone oestrogen. Some women have more oestrogens produced from Olive ovary than normal for femaleness (blondes especially) plus more cycles and preparation for pregnancy than others. Oestrogen has made it into the water supply (from oral contraceptives) and is affecting Speedo sperm counts as well as changing sex organs in some animals e.g. polar bears and fish. Some people believe this hormone is caught up in the water cycle and it even rains oestrogen. The balance of the hormones testosterone and oestrogen can make humans more feminine in actions and appearance and women more masculine. It is often said that as men and women age some reversal is seen i.e. older men become more caring and older women more aggressive?

Oestrogen protects Happy heart in women so this is possibly a reason why women tend to live longer. High levels of oestrogen are thought to be linked to breast cancer especially where it is stored in fat. Some oestrogen replacement therapies have been implicated. Exercising to reduce excess fat can reduce oestrogen storage. Chocolate also is believed to have a similar effect as Oestrogen.

Men and women also produce pheromones (scent hormones). Women can detect musk at a very low dilution undetectable by males. In a bull elephant at musth which it goes through two months every year it excretes a cocktail of chemicals in mucus from its cheeks and more than three hundred litres per day of green smelly urine. The smell can be picked up half a mile away by cow elephants. Some animals leave scent on their footprints. They say that the vanilla smell in France is feminine and elegant, lavender is masculine and stimulating.

Certain smells can affect women and certain smells affect men. Some shops use the relevant perfumes/smells to encourage shopping, change human behaviour and to influence choices. Urine and faeces can also contain specific chemicals to mark territory. Dogs wee on lampposts which can rot. Hippos defecate, backwards in rivers, dispersing the poo and its scent with their tails. Damage or loss of function of hormones are important especially during development Too much growth hormone can result in giants in adults, big hands, feet, skull and face and too little growth hormone can produce dwarfs. Growth hormone comes from Thelma thyroid gland and contains one or two atoms of iodine.

Another member of the hormone team which is wrapped around the windpipe. This hormone is produced in one or two pulses per day as well as when you are sleeping. More is released in children which with growth factors in the Lily liver initiates cell division so you get bigger.

Children grow quicker in the spring. Your Noddy nose and Ewan ears never stop growing. Ewan ears grow 0.01 inches in length every year. The smallest person's height was 0.4 metres but 0.6 metres is more common and the tallest was 2.36 metres. In Elizabethan times the average height was 1.65 metres nowadays this is 1.78 metres though this is not always hormone related. An example of this is seen in some children where protein intake is increased resulting in giant's big hands skull and face. Not all dwarfs are linked to hormone deficiencies. Primordial dwarfs are rare and do not have children. At age sixteen they may only be two feet high. Some parents with seriously handicapped children have used hormone treatment to keep their children as small children and a manageable size to transport around.

Near to Thelma thyroid gland are four Patricia parathyroid glands each only 6 millimetres across. The hormones from these are regarded as being responsible for gill development, which in humans are now inside you and are used for calcium regulation.

Theodore thymus gland is often sold as sweetbreads and is found at the base of the neck. Their role is to produce hormones that help with maturation of Winston white blood T cells involved in immunity especially in the young. Theodore thymus shrinks with age being replaced by fatty tissue. It also stimulates some Pippa pituitary gland hormones.

Blood sugar is regulated by the hormone insulin produced by Peter pancreas. Insulin helps convert too high levels of glucose in the blood into a larger sugar called glycogen which is stored in Mollie muscles and Lily liver. Insulin also helps other body cells to use glucose to reduce the amount in the blood. On the other hand if the blood sugar is low, then another hormone in Peter pancreas called glucagon (glucose gone) beaks down the glycogen stored in Lily liver and your Mollie muscles to glucose so it can travel round the body with me to where energy is needed. Peter pancreas therefore acts as a sugar monitor but also the hypothalamus in Brian brain is involved as well. Another hormone somatostatin, in Peter pancreas controls the glucose glycogen switch so that time is given for blood levels of glucose to level out.

There are many other hormones in the body, oxytocin being the one responsible for feelings by children towards their mother.

Ageing and Hormones

With ageing hormone production lessens also Brian brain regulating mechanisms are slower to respond. Older people can have higher metabolism, and an overactive Thelma thyroid can produce more thyroxin hormone. It is said that this also causes older people to be more moody, not settling down and finding faults in many things. Hence the term 'Grumpy old men and women' and the portrayal as the battle axe of a mother-in-law.

Tony testicles shrink when old so less testosterone is produced. Perhaps that is why some men become like fussy old women, their female hormones taking over and their breasts enlarging. Women also go through the menopause where oestrogen levels drop as Elsie eggs are no longer produced. Beer high in hops has a lot of oestrogen so men can develop big breasts and bellies by drinking a lot.

10 NELLIE NERVE NEGOTIATES WITH BRIAN BRAIN AND LOOKS FOR THE SIXTH SENSE

Information is exchanged within the body through hormones but a faster route is the nervous system. Hormone control is slow and also takes longer to stop due to the need for chemical breakdown. The body arrangement in animals and humans evolved from segmentation and bilateral symmetry. The rear was specialised for locomotion as a tail. The front of the body evolved for feeding and some limbs developed into jaws and feeding mechanisms. A concentration of nervous tissue which controlled the rear end and limbs developed also in the front of the body. As children you have development milestones. You have a predetermined biological clock for programmed development, usually the head. Brian brain develops then the organs along the midline.

Every cell in your body is electric. You are mostly water with salts which are good conductors. Nellie nerve cells can change anything into electrical energy, heat, light, pressure, sound etc. etc. The nervous system is electrical with Nellie nerves like wires. Brian brain produces enough electricity to power a toy train. At rest you can radiate one hundred watts but a shock of over three watts can stop Happy heart. An electric eel in comparison can generate a high voltage (four hundred volts) for defence or detection of prey. Some people can for short periods increase electrical charge on certain parts of their body and transfer shocks as well as attracting metal objects.

Most animals have a nervous system. Some have a neural net which in primitive animals adapts 'learns' from different inputs stimuli, from outside and from the intensity of these. The only multicellular one is a sponge. The nervous system in humans consists of the central nervous system (Brian brain and Spencer spinal cord) and the peripheral nervous system. A leech has thirty two Brian brains, a silk moth eleven and a starfish none. The sea squirt has one but eats it when it settles down on a rock for life.

In humans Brian brain is protected by the skull. One man had a groove in his skull which held a candle useful for reading. He was called the human lantern. You start losing Brian brain cells from a young age and you cannot regenerate them like a canary (I always said bird Brian brains were the best). Crows are the brainiest birds with high IQ. The dumbest domesticated animal is believed to be the turkey. One Nellie nerve cell in Brian brain can have twenty five thousand connections which means the number of messages are enormous. Messages from Brian brain can travel at four hundred and thirty two kilometres per hour. and be processed in milliseconds. The total length of Nellie nerve fibres are one hundred and eighty thousand kilometres. Neuroglial cells form the Nellie nerve glue keep them in place and help with supplying them with nutrients.

Brian brain today (pink in colour with grey matter mainly on the outside as seen in the inside of Spencer spinal cord). It is the heaviest organ in the body, two percent of the body weight, (one thousand four hundred grams. It is slightly heavier in men than in women. The chimp's Brian brain weighs four hundred and twenty grams. The human Brian brain is three times the size which probably arose after grasslands were replaced by forest and you became bipedal and had a better diet which included meat. Five weeks after conception Brian brain develops and reflex actions are seen at eight weeks. In the first years of life the size of Brian brain triples. Only at seventeen does Brian brain reach its full size. If born without a Brian brain a human baby can live eleven hours but you can, however, survive with only half a Brian brain. Dinosaurs had Billy bones greater than 1.8 metres but a Brian brain smaller than a fist.

Brian brain is eighty to eighty five percent water the same as Mollie muscle but with less protein and twice as much fat. It is sixty percent white matter and forty percent grey and uses more oxygen than other parts of the body. It also has over a hundred billion neurons (Nellie nerve cells), and around 7 million can be used per day. A bee only has nine hundred Nellie nerve cells. Brian brain uses five times more energy than the body organs. It can take in over eleven million pieces of information each second but may only select 40 pieces. It can process eighty six million pieces of information in a day and can store over four terra bytes of information. Litsz, the composer, could remember twenty –one three hour concerts .It needs a good supply of oxygen to function and burn up the glucose from the 0.85 litres of blood that passes through your Brian brain. Chop your head off and you are conscious for twenty seconds.

At the top of Spencer spinal cord is Brian brain stem. This is your primitive autopilot, reptilian in character mainly used for basic functions. These include movement, heartbeat, blood pressure, respiration and even the swallowing reflex. These are the things you don't have to think about to keep the body ticking over. This saves energy by requiring little thought.

At the back of Brian brain is the Cedric cerebellum which is cauliflower like and responsible for balance and posture.

This stores learnt motor programmes and more practice produces better fine tuning actions which become automatic. This is the part that is affected by alcohol and may let you down when you are asked by police to walk along a white line. Some people are born and can survive without a Cedric cerebellum. The gap replaced with cerebral fluid. Lack of Cedric cerebellum can affect voluntary movements and speech.

Chelsea cerebrum is the most modern part of Brian brain divided into two hemispheres which control opposite sides of the body which gives you intelligence and consciousness. The left side of Brian brain deals with logic and the right with imagination. Spread out it would be the size of a pillow case. The more wrinkles in the cortex they say the more intelligent you are.

.

On the outer surface of Brian brain is the crinkled cortex. (Composed of six layers of cells called grey matter) This crinkling enabled humans to develop their motor skills. Expanded it would be as large as a bedspread (two thousand five hundred square centimetres). An elephant's cortex is six thousand three hundred square centimetres while a cat is eighty three square centimetres. (Shrew 0.8, rat 6.0 and killer whale a large seven thousand four hundred square centimetres). The cortex can be thought of as an empty page which you fill up through childhood and life experiences. It has made you what you are giving you skills and intelligence. Regions on the right can control or affect the left side of your body and vice versa. Vision is the same but also pictured upside down but inverted to the right way by your Brian brain so as to see things right way up. It is said there is a daredevil gene which affects a protein in Brian brain which makes you more daring.

In teenagers the frontal lobe of Brian brain is not fully developed so they tend to take risks and are more emotional. They therefore rely more on these primitive urges before others are developed which are responsible for reasoning. In the past it was thought that if the mother thought beautiful thoughts while pregnant that they would produce a beautiful child.

The front of Brian brain is for thinking and learning and the back is for vision. An arch in the middle is for motor responses (movement) and behind this is another arch like area, the sensory region. The left side is for logic, maths, and language and the right, creativity, art and music. In women Brian brain is used more for language and speech to relieve problems that's why women talk more. Nellie nerves from sense organs send messages to Brian brain. Brian brain can distinguish one drop of lemon juice in one hundred and twenty nine thousand drops.

Brian brain and Spencer spinal cord is bathed in five hundred millilitres of cerebrospinal fluid which is produced every day by chambers and ventricles in Brian brain, It's mostly water but with salt, protein and glucose and it acts as a cushion. Brian brain is only linked with the skull by blood vessels so can easily be damaged so needs to avoid a hard bash. Brian brain itself feels no pain but the membranes around it have Nellie nerves and can transmit pain. Your Brian brain will start to die if it is without blood for eight seconds. Renewal of Brian brain cells is not common. Stem cells can sometimes renew if injured.

Some Brian brain areas are reduced compared to other animals e.g. hearing range. Grey matter in Brian brain and Spencer spinal cord has all the nuclei of Nellie nerve cells while white matter is the long wire like strands, with the Mickey myelin sheaths the axons of the Nellie nerve cells. Spencer spinal cord which extends down from Brian brain is forty five centimetres long and stops growing when you are five years old. It is a few centimetres shorter in females and contains about one billion neurons. It does not reach right down to the base of the spine. It has a hollow centre with an H shaped cavity which contains Spencer spinal fluid which contains nutrients and minerals.

Seventy two kilometres of Nellie nerves form most of the rest of the nervous system. These are like long insulated wires with tentacle like extensions at each end to spread out and receive the messages and form linkages. There are over two hundred different kinds of Nellie nerves. In Charlie chest is the solar plexus where many Nellie nerves network. Hitting this spot can cause paralysis. Nellie nerves like Mollie muscles have an all or nothing response i.e. will send a message or will not because of not having enough charge to send one. To make the Nellie nerve work and feed it must have a long cell body in it. In a motor Nellie nerve the cell body is at one end, but in a sensory Nellie nerve it is in the middle as a sense organ has to be at one end. There are also connecting Nellie nerves where the cell body is covered in tentacles which connect to other Nellie nerves. This one is found in Brian brain as there are a lot of short connections needed there to send messages.

One can think of the sensory Nellie nerve cell also as being a single cell which has been pulled out to the left and right whereas the motor cell has only been pulled from one side. Nellie nerves are covered in a sheath of fat called fatty myelin for insulation but this is not brilliant. Nellie nerve impulses may only last around 0.001 second. The thicker the Nellie nerve the thicker the sheath needed. This is seen in worms and octopuses where the thicker Nellie nerves allow messages to travel ten times as fast. The fastest in human is one hundred and twenty metres per second (electricity is three hundred thousand kilometres per second). Just like wires, if the Nellie nerves touch each other, then the electrical current would be disrupted and the final signal current would not travel or would be reduced. Bundles of fibres are covered by a protective tissue layer and then another layer encloses this which includes blood vessels.

Some fibres are not covered in fat (myelin). Examples include skeletal, cardiac and smooth Mollie muscle. Often there is no myelin in places where Nellie nerves are very small. Mollie muscle conduction is slower. Pain conduction is slower than touch. Plants can also respond to stimuli but it's not nervous, the fastest response is for dogwood at 0.5 milliseconds.

The longest Nellie nerves are in the leg. One of these is the sciatic Nellie nerve (Latin - pain in the thigh) which is really two Nellie nerves joined together. This is over a metre long and runs from Spencer spinal cord to your big toe. The giraffe has a Nellie nerve running 5 metres toe to neck and in a killer whale Nellie nerves may be many metres long. The smallest Nellie nerve cells are in the Cedric cerebellum of your Brian brain and are only 0.005 millimetres diameter. Grooves in Billy bones take Nellie nerves so that Mollie muscles don't squash them. Nellie nerves can join together in the body to produce bundles called ganglia.

The nervous system has two opposing systems like a door to open and close, to start and then reduce or stop an action within the body. These are parasympathetic and sympathetic nervous system. Parasympathetic Nellie nerves emerge from your Brian brain stem and Spencer spinal cord near your tail. Sympathetic Nellie nerves emerge from your Spencer spinal cord at Charlie chest and back and the effects last longer.

The nervous system has also a safety primitive sudden reaction found not only in humans but in most animals which is called the Reflex Action. It is for your survival so that you do not damage your bodies or lose a limb. In animals a sudden massive stimulus on a sense organ (bright light in Ellie eyes, loud sound in Ewan ears, pain, touch, pressure in or on Slinky skin, bitter, sweet, sour, on your tongue and pain etc.) causes a Nellie nerve impulse to travel to the spine first and back to a Mollie muscle for instant response.

Examples of the reflex arc are a sudden reaction to a pin in a finger or a hot iron on the hand. Messages zoom from the sense organ and travel along sensory Nellie nerves to a relay Nellie nerve in the spine and back by a motor Nellie nerve to the Mollie muscle causing you to react. Note Brian brain here is not initially involved a shorter route being taken. The stimulus sill does also go to Brian brain to check whether to initiate further action i.e. remove or get away from the object. Two processes are involved, motion and Mollie muscle inhibition.

Another primitive reflex is the grasp reflex seen in a baby. This is normally lost in a few days but can last for several months, then develop again at seven-eight months. Another reflex the cling reflex, occurs when babies fall backwards and throw out their arms. They put arms and legs together and this may be a remnant from your monkey past (apes also have this) for hanging onto mother's fur while climbing trees or running away from predators. Blushing in children also develops as a reflex action between two-four years of age.

As a red blood cell I travel around Brian brain and many sense organs. Fifteen percent of blood leaving from Happy heart goes to Brian brain. Many receptors in your bodies are sense organs. Receptors may vary in the fields they detect over and also their adaptation times. Some animals e.g. hammer head sharks can detect voltage less than a billionth of a volt. Sense organs together with receptors in Brian brain, maintain your body at a constant level (homeostasis). The largest sensory receptor is Slinky skin detecting different kinds of stimuli. The hands have a lot of Slinky skin receptors with Nellie nerve endings of two hundred per square centimetres compared to finger tips which many more, two thousand five hundred per two square centimetres.

Sense organs help to show Brian brain what has happened. Baroreceptors in Happy heart monitor pressure of blood. Thermoreceptors (also inside the body, Brian brain and Spencer spinal cord) feed back to Brian brain constantly. Reduce the body temperature then a bath will induce sleep. If your body temperature falls by 4 ° centigrade you become unconscious. Receptors can adapt by constant emerging in cold water over several years and some swimmers benefit from this when swimming in winter. Beethoven dipped his hand in cold water before composing.

There are also chemoreceptors in Brian brain and body which monitor metabolites, pH, carbon dioxide, oxygen and glucose levels and stretch receptors which monitor tension, stretch pressure and position. Pressure receptors monitor touch. Touch receptors respond faster than pain receptors. You cannot tickle yourself and cause laughter as you are already aware of the stimulus. One can read faster with the tongue than Ellie eyes and can even read Braille with your lips.

For touch stimulus pressure on Slinky skin has to be 0.01 millimetres deep to register. Most sensitive are the lips, Moby mouth, sole of feet and elbows, also at the base of your spine. Buttocks are less sensitive. Females are more sensitive to touch than males. Finger tips sensation is reduced during female menstruation, but during pregnancy all the women's senses are heightened which could be protective against danger. Some pressure on Slinky skin can cause headaches as seen with a tight bra on the cervical Nellie nerve. Touch receptors in Moby mouth give information on texture. The itching sensation may be a remnant of when you scratched to remove fleas and lice which could cause infection.

Pain receptors are the most abundant. Acupuncture dulls pain by affecting the passage of Nellie nerve impulses. It is believed that firewalkers have a gene absent for a protein which allows them not to feel pain. Thermal pain starts at 45 ° centigrade. Covering burns with cling film then ice on top slows down damage and helps repair. Human Sally saliva also contains a pain killer, opiorphin, which is said to be six times more effective than morphine. Thresholds of pain can vary for different professions. For soldiers the pain threshold is higher than for a desk worker. The best human pin cushion had thirty two needles in the arm for thirty one hours.

It is thought that when you are born all your senses such as taste, vision and hearing initially overlap. As you grow these separate but for some people these can still remain where people link numbers with colour or taste. These people can use the presence of one sense and transfer this sense to another in Brian brain e.g. smell colours. This is called synaesthesia. Some people, deaf at birth, develop other senses to compensate and can recognise people by smells. You cannot taste unless food is moist and is dissolved in Sally saliva. Taste receptors are on your tongue but not many are found in the centre. They can be replaced every week. Babies have more taste buds than adults and also have them inside their cheeks. You still have them on the roof of your Moby mouth. Females have more taste buds than males. Taste buds are more sensitive in the mornings. Sweetness, it is believed, initiates suckling, which may be the reason why under stress you eat sweet things more.

You have a sweet tip to your tongue, salty either side, and behind this is sour. It is said that some people can detect one gram of salt in five hundred litres of water. The middle and back of your tongue detect bitterness, hot and cold. Drinks can influence your taste. Certain wines and beers can affect the fats in Moby mouth and affect flavours. Carbon dioxide in drinks can clean your palate. Some drinks or beers, especially if malted, can taste of other flavours such as chocolate or coffee although they do not contain these ingredients.

Olfactory (smell) receptors in your Noddy nose register distance and chemical communication. The smell of food is important in food preference and enjoyability. Slugs have 4 Noddy noses. You can adapt to bad smells eventually and do not notice them e.g. living next to sewage treatment works. After birth immediately the sense of smell deteriorates. The tongue and inside of Noddy nose needs to be wet to detect smells. The total number of receptor cells for smell in a human is only twelve million compared to a blood hound's four billion over an area thirty times as big. It is believed humans lost their sense of smell compared to dogs, in favour of sight and sound. Ten million of these smell receptors in humans are lost every thirty days and they don't function well when covered in mucous i.e. when you have a cold.

Smell is ten thousand times as sensitive as taste. Women are known to have a better sense of smell than men that is if they don't have a cold. However they have less of a response to the smell of coffee. They also, unlike men, can smell musk the lowest concentration is 0.00004 milligrams per litre of air or one drop of musk in the Albert Hall. In the middle of their menstrual cycle the sense of smell increases. A baby can even smell a mother's milk plus body odour and will turn towards her breast even in the dark.

Noddy nose can distinguish over ten thousand chemicals/fifty thousand different smells. and you can learn to distinguish the differences just as you can for taste. Odours include acidic, burnt, pungent and fragrant. It is said that smelling green apples or bananas helps you lose weight and grapefruit makes you feel younger. There are also smells which increase sexual attraction in men. Cinnamon or pumpkin causes sexual arousal it is said in men while women respond to liquorice and cucumber. The human sense of smell is constantly decreasing over the years. The worst smell is believed to be onions in poo. Even plants are believed to give out distress calls and example being the smell of freshly mown grass.

Sight has developed from simple Ellie eye spots which form in a depression which then developed a lens over it. Human male's Ellie eyes are slightly bigger than females. Some people can pop their Ellie eyes out to twelve mm from their sockets. As in most predators you have binocular vision. Lizards can have a third eye with a lens. Clams have thirty five blue Ellie eyes. Only lemurs and you humans in mammals have blue Ellie eyes. The human lens can change shape over one hundred thousand times per day depending on activity and in a lifetime may process over thirty million images. The size of the pupil at the centre of the eye it is said decreases with age.
Having two Ellie eyes gives depth. The blind spot is not seen as Brian brain compensates. Lateral vision with Ellie eyes on both sides, like a donkey means they cannot see things directly in front e.g. a whale may not see a boat coming. Although you cannot see four legs at a time like a donkey you, like owls can judge depth. and are more sensitive to the slightest stimulus. Some animals e.g. seals and members of the cat family have to see in poor conditions or at night and have a crystalline layer called a tapetum at the back of Ellie eye. This reflects light back so

they have a second chance to make out an object. That's why at night you see cats Ellie eyes shine. Nocturnal animals also have a reflective layer so can get a second picture which helps them detect prey. Their Ellie eyes are often larger, as seen with owls and deep sea fish, so as to capture as much light as possible at the back of the eye on the retina. You have what are called rods (black and white detecting cells) which are useful for seeing better in the dark. Owls have more rods than you have.

You also have cones, (colour detecting cells). Rods are situated mainly on the rim of the retina while cones are near the centre where vision is best. The retina also has five to six million cones and one hundred and twenty to one hundred and forty million rods. The more receptors that are present the sharper the vision over long distances. It needs a photon (packet) of light to excite a rod but one hundred photons to excite a cone. This means you can see colours better in bright light.

Cones rely on pigments and overlapping of primary colours, three colours in humans green, blue and red. You can though detect ten million different shades of colour. No colour is naturally blue, bluebells are really purple. Some animals, we believe, see more. Some primates only see green and blue. Some animals are also more sensitive to one colour. Cats have poor developed colour vision but dogs and horses also have some colour vision. Bulls are colour blind so don't see red so waving a red cloak by bull fighters may not be the full reason for attack. It is said you can detect over eight million colours. Light stimulates and interacts with your pigment in retinal cells in Ellie eye sending Nellie nerve impulses. This is detected in two hundred femtoseconds (one is a millionth of a billionth of a second). Different wavelengths of light give the different colours but you only see a few colours. You have some filter against ultra violet light but some animals see UV and distinguish objects or predators. Some snakes can see infra-red light but this is detected by a separate organ on their heads. If all the light is absorbed by the object then you see black, if all is reflected you see white colours. In a black hole no light can escape so it's black. There is a belief that there are people who can see ailments in humans, a type of ray vision where damaged organs give out a signal.

Pressure on Ellie eyeball can excite the retina and you can see phosphorescence. Neurons inside the eyes fire even in the dark to give star like flashes.

Seven principal Ellie eye movements are believed to exist. One of these involves looking straight ahead which gives you maximum information. It is believed movements down and to the left relate to something you associate with something previously e.g. taste or smell. Ellie eye movements also go in an opposite direction for left handed compared to right handed individuals. What you normally see may not be the real picture in time as the optic Nellie nerve receives millions of messages which are passed to Brian brain. You see around sixty images per second other animals can see them faster.

Sub minimal messages can be given by watching fast images which are too fast to comprehend but Brian brain can pick these up and these may influence actions or choices unwittingly. Magicians will rely on distraction, you focussing on one action at a time. You can blink fifteen times per minute or twenty five thousand times a day, especially when you are talking which is a lot of stimulation!

Hearing involves small Billy bones respond to movements of your Ewan ear drum magnifying the changes in movement. You can also hear with Billy bones in the jaw and skull. This is made use of where sound can be transmitted through bones using a transducer instead of earphones. The sound that you hear of your own voices passes through liquid and bone and so will be different from what others hear. If born without Ewan ear openings you can listen through your Moby mouth. Beethoven who was deaf used a stick in his Moby mouth and the other end placed on a piano while he composed. Hearing, it is said, is at its best at 10 years old. The piano has seventy six octaves 440-3520 Hz.

Humans can hear frequencies from 20-20,000 Hz. The human Ewan ear is most sensitive to 1000-4000 Hz. A lot of animals hear higher frequencies. Dogs cannot hear low notes but can hear high pitched notes the highest up to 46,000Hz. They are believed to be good detectors of disasters compared to other animals. Rats can hear up to150, 000 Hz.

The human Ewan ear is sensitive to units of sound intensity in decibels. The greater the amplitude the higher the frequency. A ticking watch is 20 dB. At 35 dB this sound becomes annoying. At 55 dB this causes stress and at 90 dB (e.g. a pneumatic drill) this can cause pain. At 120 dB, as does 90 dB for a long time, this can cause damage. Sound can affect blood pressure and produce hormonal changes. Living in a city where sound is a problem or near factories or airports can increase stress levels and initiate heart problems.

Some animals can swivel their Ewan ears towards the sound. The Chinese believe the longer Ewan ears you have the longer you will live. The denser the medium that sound passes through the faster sound travels. Sound travels faster in water. Whales such as the humpback can hear 10 kilometres away and have an elaborate form and range of communication. Animals, unlike you, have an art of recognising members of their families by the sound they make.

The female emperor penguin recognises the male after several months' absence, when she returns to look after her chick. The male then has to recognise the chick's call when he returns, from among hundreds of others. You say that a human mother also can in some cases recognise the call of their baby so possibly you still have traces of this ability. It is said humans can recognise someone by the way they walk. Birds and some other animals have iron particles of magnetite built in to their skulls which help them with direction for migration to feeding or nesting sites. Another theory involves the formation of stable molecules triggered by the earth's magnetic field. You respond to magnetic fields which influence the body. Frogs can even levitate in magnetic fields.

Nellie nerves also go to Gladys glands to produce hormones for response. Some Nellie nerves go to involuntary Mollie muscles which work on their own inside the body e.g. Gillie gut. Other Nellie nerves go to voluntary Mollie muscles over which you do have control. Fewer Nellie nerves are needed if there is massive movement of Mollie muscle. But this also depends on the number of Mollie muscle fibres. Fewer Nellie nerves, plus fewer Mollie muscle fibres can perform intricate fine movements; threading a needle for example and Ellie eye closure. Some animals e.g. fish, can feel with their tails as well as with their fins.

Messages in nerve cells are electrical, produced by the movement of positive sodium ions (salt) moving into Nellie nerves. The Nellie nervous system produces electricity over the body which can build up to ten thousand volts. As this is only a small charge, the current that causes the harm is tiny. The resting potential between cells in your body varies from minus twenty to plus twenty millivolts. An electric eel can produce six hundred and fifty volts. Mental work increases Mollie muscle tone so that the metabolic rate increases. Using your Brian brain in thinking can use energy and that is why you need a lot of glucose.

The junctions between Nellie nerves and other Nellie nerves, and Mollie muscles or Nellie nerves and Gladys glands are called synapses. In Brian brain there are one hundred and fifty million in the cortex. Between Nellie nerves and other Nellie nerves or Mollie muscle, in these synapses chemicals are produced which transmit the electrical impulse. These impulses can travel over the junction as a chemical in 0.001 seconds. These chemicals called neurotransmitters which cross the gaps, when reaching the second Nellie nerve or Mollie muscle or Gladys gland produce a new impulse which is proportionate to the first. Each motor Nellie nerve i.e. to Mollie muscle or Gladys gland may have six thousand synapses. Some of these neurotransmitters may inhibit whereas others stimulate. Two systems exist for involuntary (autonomic) nervous control in the body. The systems act as switches one turns on, the other turns off the response to a stimulus.

Synapses are affected by drugs and chemicals which you take into your bodies. Nicotine (e.g. from smoking) mimics transmitters and can block the chemical message thus having a calming effect. Alcohol can completely stop production of a transmitter so less messages are sent (no wonder reactions are slower after a drink). Insecticides on the other hand prevent enzymes from breaking down the transmitter after it has sent its message. Here the result is that the message continues non-stop and if it's going to your Mollie muscles you keep twitching until your Mollie muscles seize up and no longer move. This is dangerous, especially for those Mollie muscles involved with breathing. Caffeine in coffee and some drinks boosts mental alertness. and have been used to stimulate premature babies to breathe. The use of caffeine for keeping you awake was developed by shepherds that had eaten coffee beans (awake at night) seeing sheep and putting two and two together Tryptophan is a sleep inducer found high in night drinks but also in peanut butter and banana sandwiches which helped Elvis Presley to sleep. Smoking can increase the number of receptors that respond to nicotine in Brian brain thus encouraging more smoking in the future causing a habit. Heavy meals increase blood glucose and can switch off Brian brain cells that keep you alert and awake. The neurotransmitter Dopamine can improve working memories as well as giving the feel good feeling.

The more chemical produced or the more junctions that exist will increase the impulse. The intensity/frequency (timing) of its production i.e. changes in charge, also determines how big the response will be. The number of Nellie nerves firing (i.e. responding) and the arrangement of the Nellie nerves and positions helps to determine the final strength and kind of response. Minor stimulation may not give any response at all. Maximum discharge for Slinky skin for heat and pain is at 42-45 ° centigrade. Cold pain starts at 17 ° centigrade, normal body temperature being around 37 ° centigrade.
You need pain in order to stop us damaging yourselves. There are different degrees of pain, illustrated by the types of pain caused by different insect stings. Pain is overcome by the production of endorphins, the natural pain killers which act on receptors. Brian brain feels no pain if injured but produces endorphins to reduce pain from elsewhere in the body. Placebos, which encourage endorphin production, can have the same effect. Calcium slows down Nellie nerve impulses and stabilises the response but too

little and the Nellie nerves will get excited and this can lead to Mollie muscle cramps. Potassium will also redress the chemical balance in Nellie nerves. Often bananas are eaten as a source of potassium to relieve stress. A Tens machine produces an electrical current which opens and closes the pain gates. Stress or pressure can also reduce the degree of pain.

Early on children have sensations of body movement but these are eventually taken for granted and blocked out of major thoughts. The brain always looks for and stores the unusual. Television or computer watching is believed to shape neural networks differently to that produced by physical play or reading. This may make children less receptive or motivated to some situations. Training of lots of people together at one time helps rapid learning of a skill but not retention. However it has been found spaced distribution of training gives slow learning and better retention.

You must feed Brian brain. Brian brain uses twenty percent of the calories you eat. Oxygen and sugar are essential. You can have as many as fifty thousand thoughts a day. Steady carbohydrate (sugar), especially a big one like starch (e.g. cakes and biscuits), helps to feed Brian brain better than simple sugar which does not last long. You are most alert at 10.00 hours, best coordination is at 2.30 pm and fastest reaction is at 3.30pm.

Most of how Brian brain functions is still a mystery. The amygdala part of Brian brain processes emotions such as fear and anger into memory recall. However, you seem to have the ability to use abstraction which gives you an advantage and prepares you for threats in advance. Bad experiences produce a stronger stimulation so there are more functional connections within Brian brain and so

retention is better. and you can evaluate similar situations.
The more social the species the bigger the amygdale.
The hippocampus part of Brian brain acts like an address
book and has a time component.

.The greater number and frequencies of response help to
lay down memories (this is likely a protein). Short term
memory lasts only a few seconds but chunks last a few
minutes before being made into long term memory.
Retention is said to be best between 1000-1400 hours. (tell
that to schools and colleges). Vitamin E as found in green
vegetables, is said to help with memory retention. Skills are
often remembered short term and the first part is recalled
but not the middle. Babies produce language (first words)
from eight to twenty three, months Women use twenty
thousand words a day while men only use seven thousand.

Nellie nerve cells transmit energy which can leave or enter
the body as charge (voltage). Fillings in teeth in the mouth
can conduct this body electricity. You are not batteries and
do not store charge. Some people produce more charge
and static electricity than others and they may be
responsible for healing so that charge is lost when one
makes contact with another person. Even when kissing
one can start to drain the energy from the other person.
Some people can eventually, if very ill, emit excess charge
as light or can exhibit magnetic properties. You emit
energy as infra-red, as much as a one thousand watt light
bulb in some cases. You also can gain energy when an
animal jumps onto your lap and especially by stroking
animals. This helps you increase charge which increases
your wellbeing. Animals are encouraged in some old
people's homes so as to help them recover from illness.
Illness could be reflected by low charge or in some cases
very high charge where organs dysfunction.

As well as animals and other humans everything you touch can alter your charge. You can download charge e.g. static from pointed metal objects such as the corner of a desk, a handle of a pushchair or shopping trolley. An example where a visible version of static charge is produced by using millions of volts produced using a Tesla coil. A side show exists where madam Electra has two hundred and seventy thousand volts travel harmlessly through her body. She can light up bulbs, fluorescent tubes and emit sparks from the end of her fork in her Moby mouth. However care has to be taken as the body cannot detect current above twenty kHz so by touching the emitted sparks the current can do unknown damage inside the body. All the organs in your body also produce a voltage, Brian brain and Happy heart especially.

Even when you are dead electrical waves in Brian brain can last thirty seven hours. During a storm a charge can enter your bodies by lightning and go down to the positively charged ground through your legs. The voltage can be as much as three hundred thousand volts and the current can be very high but only in an instant. Current can move over the outside of the body and can cause burns but can also stop your Happy heart or Lucky lungs working by affecting control centres in Brian brain.

The front part of the hypothalamus is responsible, it is believed, for driving body biorhythms and setting/resetting rhythms of peripheral oscillators. Every atom in your body oscillates at ten to sixteen Hertz. Happy heart oscillates at ten Hertz or once per second.
Most animals some time of day are affected by dawn and dusk. The insect cicada emerges after of thirteen or seventeen years. Cycles each day of light and dark dictate physiological cycles. Sunrise speeds up the clock, sunset slows it down. If humans are blind then this is no good, but birds, reptiles and amphibians have extra ocular

photoceptors which can still detect these changes. In the body, especially in organs there are clocks for everything which it is believed are linked to a main clock which is reset by daily and other rhythms. There are clocks for protein synthesis for the production of many molecules, hormones, enzymes etc. Brian brain tracks the degree of light through Ellie eye. You also have a third Ellie eye it is said behind Noddy nose which still responds through Slinky skin to light and dark.

Rhythms of life are affected by gravity. Higher up a block of flats etc. you age slower than lower down. Certain times of day are believed to have common occurrence in the body. At 3-6 pm your blood pressure drops. Between two and four in the morning old people die and between three and five in the morning most births occur when temperature and pain is at its lowest. From the age of 3 to 21 years old humans get up later than at other times in life. Energy levels are lower between 1 and 3 pm, being lowest at 2pm. Depression is said to occur most on a Wednesday.

You can spend six years of your life dreaming. On average you can have as many as four dreams per night. Brian brain is more active at night and brain waves occur more often between 12 noon and 3pm. After ninety minutes into sleep more waves are seen when Brian brain is very active The Tibetan monks learn how to remove negativity by affecting Brian brain's gamma waves. During sleep you can twitch (sudden Mollie muscle contractions) and may wake up suddenly.
You need sleep to rest, repair Mollie muscles and replace cells. You need five hours of deep sleep per night Babies often need twice as much sleep than adults. They sleep half the day if lucky. Old people also sleep less. When tired the average person falls asleep in seven minutes. Brian brain has four main ranges of Brian brain wave frequencies as measured by EEG (an electro encephalogram). These are

0.5-4, DELTA/ 4-7 THETA/8-13, ALPHA, and 13-30
BETA.

Sleep has four cycles of NREM (non-rapid eye
movement).These are interspersed by individual cycles of
REM, (rapid Ellie eye movement). The NREM cycles at
each stage produce a deeper sleep. You dream in the REM,
your Ellie eyes move side to side under your eyelid and
dreams are more real, based on experiences. If you dream
before you wake you can remember more but they are
often unrealistic. Your long term memory does not work
when dreaming so the last dream is what you will
remember. Sometimes the start of an event in a dream or
noise will wake you up. During this time your Happy heart
beat and blood pressure can increase but then starts to
decrease and your Mollie muscles can be at their weakest
like paralysis. NREM dreams are more intense and bizarre.
Brian brain is very active in NREM. Body temperature
drops as sleep deepens, Brian brain waves are slower and
in deep sleep they are even slower when Brian brain
temperature drops to its lowest. Cortisols (steroid
hormones) get you up in the morning and make you ready
to deal with stress etc.

With the onset of winter and shorter days these have a
worse effect than longer days on you. With reduced
daylight from October -March, Melatonin is released more.
Melatonin is a hormone produced in response to a
biological clock triggered by light levels on the retina and
detected by the pineal gland in Brian brain. Melatonin is
inhibited by light and is important for sleep patterns. It is
secreted at 0400-0600 am and is maximum at 9.0 pm and
production stops at 7.30 am but can change with
environmental and physiological circumstances. Some
cows are milked at this time and the milk sold to humans
with melatonin in as a nightcap. You lie down when
sleeping but animals such as cows and horses stand up.

MIKE PEARCE

The effect of seasonal light changes can also produce the
Seasonal affective disorder SAD. This often starts in
winter as the days become shorter. It is believed to affect
levels of melatonin and serotonin in Brian brain causing
bouts of depression, lack of interest and more inactivity.
Some people use day light lamps to prevent SAD at home
or even in the office. People can affect their cycles by
doing night shifts. You can keep your Ellie eye closed for
ninety days or train yourself to lie on nails or stand on
pillars for days. You can go without sleep for over eleven
days but this can be dangerous and lead to death. Some
people unfortunately have narcolepsy where they fall
asleep during the day at any time. This can happen many
times during the day. Catalepsis also exists where you can
lose Mollie muscle control and nap in the day with dreams.

Some animals may have other circadian cycles which relate
to day length. You may have lost these or never had them.
The lack of seasons in the tropics where you are supposed
to have originated, could be why you never fully developed
these extra cycles. You have a twenty four hour cycle for
many of your body systems. Shifts of these cycles can
cause illness depression and death at a younger age. The
lunar cycle also affects behaviour. Violence increases
around a full moon and also activity of the mentally
unstable. Disorders e.g. asthma are affected by lunar cycles
as well as reproductive cycles You all have biorhythms,
the average is twenty eight days for one physical cycle.
Emotional cycles are supposed to be twenty three days and
intellectual cycles thirty three days.

Body temperature (including that of Brian brain) is
minimum in the early morning but usually at a maximum
in late afternoon. Brian brain temperature is maximum at
midday. Exercise can raise the body temperature to as

138

much as 40 ° centigrade. Women can also know when they have ovulated as their temperature rises.

In warm conditions the human body adjusts temperature by feedback from Brian brain. This can cause sweating and blood vessel expansion in Slinky skin so that we blood cells and plasma move faster. In colder conditions less blood goes to Slinky skin and temperature is raised by shivering and an increase in metabolism or the search for warmth.

Nervous output and sensory sensations can become adapted by repeated stimulation so that stimulation is no longer registered and the effect reduced. Examples are in older people where the body's own pain killers help to reduce pain. Pain is evaluated with the pain experienced previously. Also, if exposed to cold now and then, Mollie muscle shivering is less as you are more tolerant. In very cold conditions metabolism also adapts and increases to keep you warm. These effects must have been important to primitive man who survived an ice age and also is effective for those taking a dip in the sea every day of the year whatever the weather.

Baby reindeer can sleep in snow, but die if wet. Reindeer get antifreeze from some of the mosses they eat. The set point of temperature in the hypothalamus is altered by toxins, produced by pathogens or substances released from Winston white cells. This causes a rise in body temperature resulting in fever and sweating. The higher the temperature rises the more is the chance to kill the pathogens. More blood to Slinky skin is therefore cooled. If the temperature falls, thyroxin and adrenaline is released increasing blood flow and Mollie muscle activity.

Nervous system —diseases, disorders and injuries

Body pain can be in another part of the body from which it originates Some parts of the body were linked as an embryo e.g. the left arm has the same segments as Charlie the ches.t Internal pain is felt in those areas which are related and developed together in your ancestral segmentation. This means that a pain in Happy heart also felt in the neck and arms is often used as an indication of the onset of a Happy heart attack Barometric pressure falling before a storm can cause joint tissue to swell for those with arthritis it is believed. Increased fluid movement in the body can cause pain

Continual pressure on Slinky skin can cause pins and needles where Nellie nerves are squashed. However, light touch such as stroking opens and shuts pathways in Brian brain. Oxytocin is released which is relaxing. Floaters tiny blood cells in the liquid parts of the eye can often be seen by many people against a white or misty background. Brian brain can learn to ignore these; however numerous floaters can indicate retinal damage. The cortex for vision can be adapted for use by tougher senses in a blind person. A person blind in one Ellie eye only can lose a fifth of their normal sight. Some people have synaesthesia where sounds, taste, numbers and smell can be associated with colours.

Effects on the brain can include too much information coming in. This can wear a person down an extreme example being autism. Caffeine stimulates Happy heart and your respiratory system keeping you alert. Any shock or trauma at any age can have various effects and can even change the colour of your Harry hair. Fear of execution can turn black Harry hairs white overnight.

Spina bifida is where Spencer spinal cord in the baby is
exposed as Billy bones are not well developed. This is high
up on the spine where you get problems such as leg
weakness, numbness, lower limb paralysis and
incontinence. This is normally dealt with at birth. Use of
folic acid during pregnancy can prevent spina bifida.
Hydrocephalus is the obstruction of fluid flow which
results in enlargement of the head in children causing
thinning of Brian brain. This may be linked to Spina bifida.
Removing the cause or draining off fluid is required.
Can also get cyclopic infants with one Ellie eye where the
forebrains have fused.

Cerebral palsy is a form of Brian brain damage in children
caused by Veronica virus which affects development. CJD-
(mad cow disease) is due to a prion protein which replaces
the normal proteins in Brian brain. This results in holes in
Brian brain tissue so signals are affected. CJD can be
caught from eating infected spinal products, also from
corneal grafts. It can lead to dementia but can take several
years to develop

Meningitis can be Veronica viral or Bronwyn bacterial and
results in inflammation of the membrane over Brian brain.
Symptoms include a raised temperature, stiff neck,
headache and vomiting, floppiness and fits. Cold hands
and the fact that reddish spots don't fade when Slinky skin
is pressed with a glass are also symptoms. Delay is bad,
often resulting in disability or death especially from the
viral form, if not spotted. The source can be hard to
identify.

Many people have obsessive compulsory disorders. This
can include checking things several times, putting items in
pairs or in the same place and repeating movements or
speech. Cleanliness is an example where chemicals are
overused to prevent germs. Also worrying about the house

burning down or forgetting passports, and pin numbers are other examples. OCD can make people housebound which can affect their health and social views. Where the emphasis is on one task or subject i.e. often have to do the same thing every time or can upset the person. This can be autism.

Parkinsons results from degeneration of dopamine realising neurones. More neurotransmitter acetylcholine is produced resulting in lack of co-ordination and control. Shuffling of feet and slow movement is seen and Mollie muscle tremors in fingers and shaking. Intellect though is not impaired.

Polio is caused by a virus. It invades Spencer spinal cord and increases with muscular exertion causing paralysis. The problem is serious when it reaches Lucky lungs and Mollie muscles for breathing is affected.
Rabies is a virus which builds up in the Sally salivary glands and travels to Brian brain. A bite by an animal can spread it and result in spasms of the throat Mollie muscles so lots of Sally saliva is produced, Death is from respiratory Mollie muscle spasms affecting breathing.

Stroke is due to high blood pressure. A blockage in an artery produces lack of blood to Brian brain so less oxygen is available and cells die. Paralysis is often on one side of the body linked to the opposite side of Brian brain.
Schizophrenia is where positive symptoms include false delusions and visions while negative symptoms include reduced interest in life and blankness. One needs to observe the person for a diagnosis. Antipsychotic drugs, rehabilitation and the need to protect them from harm is essential.
Migraines are often the result of low levels of serotonin. This chemical makes blood vessels suddenly narrow in Brian brain and resulting in headaches. Then they widen

again or sometimes you see zigzag flashes. Many factors can trigger them and is often linked to hormone levels especially in women and stress or even flickering lights/computers.

Disconnection can occur with areas of movement in Brian brain. This can cause Anarchic hand syndrome. This can result in fidgety hands or even in rare cases cause you to try and strangle yourself.
Brian brain damage can occur with lack of oxygen at birth but babies can be kept at a lower temperature to reduce this damage. In adults this can produce a catatonic state of paralysis which could last for many years but could be broken by drugs or other stimuli. A word, recollection of a sound or speech by a relative may help to take the person out of this catatonic state.

Epilepsy is where sudden electric discharges in Brian brain cause fits or seizures. There are various forms. In a grand mal you can lose consciousness and falls are a danger. Epilepsy can be controlled by drugs.
Somnambulism is sleepwalking where your area of consciousness the cortex is asleep while the area for sensory (except hearing) and movement is awake. Where sleepwalking could be dangerous, especially for children they may need to be restrained.

Head injuries are also a problem. Often after car accidents this can cause a build-up of blood and haemorrhaging. A sudden jolt by any cause can result in Brian brain being bounced against the skull so nerve cells are damaged or misshapen. Concussion is temporary paralysis of Brian brain and coma damage to Brian brain stem. Damage to a certain area of Brian brain can be less harmful than overall damage. A hit at the back of the head can transfer the impact to the front of the head affecting nerves for smell and taste.

As well as effects on body organs including Brian brain, damage and infections can also affect Nellie nerves. Huntington's chorea is genetic resulting from loss of a neurotransmitter. This causes jerky movements. Limbs and face area are affected and your personality changes. Dementia often also occurs.

A slow progressive disease which affects the motor neurons is motor neurone disease, Brian brain and Spencer spinal cord. Mollie Muscles are less stimulated and start to waste away. It may be genetic. Movement is affected, and it affects speech and swallowing but intelligence is unaffected. Communication can be through a keyboard and a voice generator and patients may require a ventilator now and then.

Multiple sclerosis - here my friends the lymphocytes and macrophages (white blood cells) attack the Mickey myelin fatty sheath on the outside of Nellie nerves so that messages are muddled. One often gets relapses with fatigue and loss of functions. Cannabis can relieve tremors.

Nellie nerve damage to peripheral Nellie nerves may be ok as others take over, but spinal Nellie nerves unlike other tissues, will not regenerate. Christopher Reeve set up a foundation to look at using stem cells, to form a bridge between broken Nellie nerves that do not grow. There have been many objections to using foetal stem cells; however your umbilical Spencer spinal cord can be frozen to provide you with stem cells later if you need them to replace organs etc. Also it is known that cells in your (Noddy nose) 'olfactory ensheaving Nellie nerves cells' are also stem cells and could help re-join Nellie nerve cells.

Cockroaches can survive several weeks without heads. Chickens with a lost head could survive eighteen months. There are places today that will freeze your head for the future so that you can be revived later when cures or other treatments are available. People who have lost limbs still think they have the phantom limbs There is still a body map on Brian brain The more functions and sense organs the area had the bigger the Brian brain area. There now are new bionic arms where circuits can be linked to Nellie nerve pathways in Charlie chest so that even sensory feelings can be registered as well as movement.

Recurrent fears have the same effect on body as many illnesses. Some staff can have sick building syndrome and have to take time off work Placebos are known to encourage the production of endorphins, the body's own pain killer, to help pain. The power of suggestion can also affect the nervous system and under hypnosis you can develop marks on Slinky skin (stigmata syndrome) and prevent wounds from bleeding. People are easily influenced by suggestion especially negative ones.

People want to be believe in all sorts of things including UFOs. Fortune tellers, spritualists, ghost hunters, winning the lottery providers cash in on this. Groups of people can be caught up in beliefs or actions. These may have not been considered if they were on their own. This can have dangerous consequences especially if the groups intend harm to others or even harm to themselves. Some behaviours are contagious and example being self-harm within a closed community.

There is also an Alice in wonderland syndrome where parts of the body appear out of proportion to the rest or sense of time is lost. The syndrome is often triggered by too much electricity in the body so blood flow to Brian brain is affected. This often seen in children who grow out of this when they teenagers. It is also known that the behaviour of mental health patients is different at full moon and also the presence of wind can bring on primitive instincts Serotonin in Brian brain can increase confidence and self-esteem so Prozac, which increases serotonin can help with depression.

Work on Brian brain is in its infancy. Already a chip can be inserted where paralysed people can think of an action to produce letters on a computer or to control actions. Functions of parts of Brian brain if damaged, can sometimes be taken over by other parts to some extent. Some people spend their life in a coma. One person was in a coma for thirty seven years and died at forty three. In the past serious behaviour problems especially in children were dealt with by lobotomies. Here an ice pick was jiggled around inside the upper part of the skull just above Ellie eye. This had serious consequences but was accepted as a treatment for many years.

Ageing and the nervous system

All of us from birth lose thousands of Brian brain cells per day. Of the one hundred billion Nellie nerve cells you only lose ten percent, Brian brain shrinks a bit and can lose 20-30 grams. At thirty five you lose seven thousand Brian brain cells a day permanently. This can be increased by taking steroids as seen with body builders. At 40-50 you can lose seventy five percent of Nellie nerve cells in the basal forebrain. At 60 there is minor memory loss. When older, poor memory results. However when old you can still improve your connections and stave off dementia and Alzheimer's. This it is believed by doing mind games or crosswords. However, your Nellie nerve impulses will still slow down so that memory cells are lost. Normal impulses are two hundred and ninety Kilometres per hour. This falls to two hundred and forty Kilometres per hour the elderly. Alzheimer's may be genetic in origin, and can be the result of abnormal protein killing cells. This is often seen when over sixty years old and results in atrophy of cortex (outer layer). Mental functioning is affected, also memory for motor skills which was laid down in Cedric cerebellum. Some possibility of slowing down the process can be made using drugs to increase the neurotransmitters.

Dementia within the ageing population is becoming very common. Research is being carried out to prevent it and support those who have it. Dementia is damage to Brian brain cells often associated with a stroke. One form is caused by spherical bodies, Lewy bodies, which develop in Brian brain. These cause confusion and visual hallucinations. Possibly this caused the change in Scrooge in Dickens Christmas carol from love of money to helping the poor. "He became as good a friend, as good a master, and as good a man, as the good old city knew".

 Sensory systems also become weaker. Balance receptors in the feet and legs become less efficient. Your sense of smell by sixty may be a problem is you do not know if you have body odour.
Hearing is the last to go. The sensory cells in the middle Ewan ear are reduced. With ageing you tend not to hear higher frequency sounds 50-8000Hz rather than normal 20-20000Hertz. By fifty years of age you can't hear sounds above 10 Kilohertz. Mollie muscles receive fewer Nellie nerve impulses so they waste away and can start deteriorating past sixty five down to twenty eight percent by eighty years of age. Fine motor skills are lost.
At eighty years old you can also lose control of Betty bladder. Also macular degeneration with loss of the visual central field is common with ageing.
When you retire you lose your biological clocks as you sleep at different times and get up later. The REM (random eye movement) portion of your sleep is smaller so you actually sleep less. This is in some cases bad for you and disorders may arise which were not present with your daily regular nine to five routines. Motivation may decrease in the elderly. However changes may be due to many other

factors which are physiological or chemical. Deprivation may be a factor, as well as environmental conditions or need for some sort of reinforcement.

Hypothermia (low temperature) effects can start at 36° centigrade. At 35 °centigrade judgement is affected. Below 30 ° centigrade you can become unconscious and at 20 ° centigrade Happy heart stops. At 10-15 ° Nellie nerve conduction stops. Cockroaches can freeze and thaw and still be alive. The hypothalamus in Brian brain is responsible for aggressive behaviour different and extremes are governed by different parts. Possibly changes in this region with age can produce changes in behaviour. Your body has created its own drugs for many years, possibly there comes a time when these are produced less or are less effective. Stress especially loss of a spouse or a marital break-up can affect the nervous sand chemical balance. These can in older people lead to premature death. Loss of a partner can lead to death in one to six months in some cases.

With sudden onset of possible death there is a sudden surge of energy in Brian brain to attempt to restart your body and get it back to normal. Brian brain death includes the death of Brian brain stem reflexes, Mollie muscle responses and respiratory response. Certain death is where no electrical signals are present in the cortex, stimulation will not increase the activity and basic Brian brain stem functions such as breathing and beating of Happy heart are not occurring.
However the human body could eventually be totally bionic supplied with implants which continue to be developed but Brian brain and nervous system will eventually die and take the human out of humanity.

11 SEX CITY, PENELOPE PLACENTA AND BABY BLUES

Primitive man, it is believed, moved from the forest to grassland. It was therefore more important for humans to form groups both for protection and for more social interaction. Child care helps survival and the development of skills by children. Hormones have been developed to encourage these partnerships. Oxytocin is the affection hormone, while serotonin is believed to be linked to obsession. Many of the causes of fights between men and women are because men are not as quick as women in picking up body language.

Tall men, it is said, are powerful, while those with smaller breasts are smarter. Finding a mate involves a complex social behaviour and in some tribes the wife is ten years older and can even carry her prospective husband on her back. In angler fish the male finds the female and remains attached to her for the rest of its life. It even uses the same blood supply. Kissing for humans can use up energy. The record number of kisses is sixty six in a minute. You can use up three calories per kiss.

Human females have the overriding urge to attract a mate. In some primitive tribes attraction is by having big bottoms and big breasts. This is now mimicked by injections of saline, implants or pads (chicken fillets). By the age of thirty breasts lose tissue and fat so bulk and size are reduced. Fat is important as where the body is fat blood supply falls. Sex hormones may be reduced and periods and Speedo sperm counts are affected. In some cultures you may marry animals e.g. a goat. Primitive animals don't need sex. A sponges can be put through a

149

mesh and each bit regenerates asexually into another sponge.

A place of fast action is sex city. In the body like Dickie diaphragm at the bottom of Lucky lungs there is a pelvic muscular Dickie diaphragm. This supports abdominal and pelvic organs.

 In the male as a red blood cell I rush into tiny caverns (corpus cavernosum) or spongy tissue (corpus spongiosum) so as to inflate Percy penis with blood and make him erect. Average Percy penis length can reach one hundred and fifty millimetres but can be as long as three hundred and fifty six millimetres. It is a good job you do not have Percy penis of a barnacle which can be twenty times the length of its body. In the blue whale it can reach three metres. Iguanas, koalas and Komodo dragons have two Percy penises.
Humans normally have one penis, but also can have two. They are usually side by side (diphalia condition) or separate if they have an extra leg. Here they urinate from one or both. Rabbits have been bred with three Percy penises. Otters and primates can have a Percy penis with a Billy bone inside. Four cloves of garlic per day are believed to increase circulation and improves erection but not your breath. Viagra also has a similar effect. In Northern Uganda some tribes attach weights at puberty to make Percy penis longer but thinner. Some can reach fifty centimetres and they often tie a knot in it to keep it out of the way. Victorious Egyptian kings collected thousands of Percy penises from their enemies. In some countries a penis is a food delicacy.

 As a Fred red cell I may also flow around Tony testes. Tony testes after birth drop into the sac called the scrotum. The scrotum of a kangaroo can be made into a purse Napoleon had two and Hitler only had one testicle.

This where Speedo sperms are the smallest cells in the body, and develop from stem cells. Chimp's Tony testes are bigger than gorilla and human Tony testes. Tony testes are eaten in a world champion eating contest every year. In deer Tony testes are taken up in the winter and dropped in summer. When cold human Mollie muscles lift Tony testes next to the body and when hot the Mollie muscles relax so Tony testes swing away from the body.

Saunas, tight trousers, over exercising, alcohol, drugs or other factors which increase the temperature of Tony testes may reduce Speedo sperm numbers. Cannabis can alter the size and shape of Speedo sperm. Alcohol also has the same effect as it affects zinc concentration which Speedo sperms need. Fewer Speedo sperm are produced in summer linked it is believed due to the presence of more daylight. Fruit fly Speedo sperm are 1000 times longer than humans. Human Speedo sperm are stored in Edward epididymis which a tightly coiled tube in the testis (6 metres long). They then flow along tubes (vasa deferens) where sugary, alkaline juice added from the seminal vesicles. A Cowper's gland secretion is also added before release. Patrick prostate gland, attached to the base of Betty bladder is only found in dogs and human males; however the equivalent still remains in women as the Skenes glands.

It takes a hundred days for a Speedo sperm to fully develop. A fifteen old boy can produce two hundred Speedo sperms a day. A healthy male produces a total secretion in a few millilitres. Pigs can produce just over half a litre. Over three to five hundred million are produced at each ejaculation. This can be ten, to thirty billion Speedo sperm per month or twice the population of the world. Man can ejaculate over seven thousand times in a lifetime equivalent to sixty four litres. The quality of Speedo sperm it is believed is increased with an interval of

6-7 days between release. A rhinoceros can ejaculate for over an hour Speedo sperm being produced every ten minutes.

Exposed Speedo sperm die in a few minutes at body temperature, but last four hours plus in the female cervix. Speedo sperm can be stored successfully for many years in straws frozen in liquid nitrogen at minus 196 ° centigrade.

Speedo sperm when released is floating in an alkaline juice which is needed because the female's Valerie vagina is acid so as to kill any germs. One could say this is a neutral partnership. However twenty percent of Speedo sperm will be abnormal with no heads, two heads, two tails or even no tails. The seminal fluid also contains fructose, a fruit sugar as found in bee honey, to give Speedo sperm a lot of energy to waggle their tails. Eating honey can enhance production of the sex hormones oestrogen and testosterone.

The temperature in the testes must be 3 degrees below body temperature for Speedo sperms to be produced. Very few Speedo sperm are lost in urine if there is no sex. There are many different methods to prevent pregnancy. You can use crocodile dung and honey like the Egyptians or moss like the Romans. Today you more commonly use condoms, creams or intrauterine devices. Low or high doses of the hormone oestrogen and progesterone can be given as patches or injections. Slow release capsules under Slinky skin can trick the reproductive cycles so that no more Elsie eggs are produced and pregnancy does not occur. However it is known for babies to be born holding inter uterine coils in their hands or imbedded in their bodies!

Reproductive cycles in women follow the human cycle of nine months. In the female of the species as a red blood cell I may travel one hundred and sixty one kilometres to reach the almond shaped Olive ovary. This only weighs 3g

(Tony testes twelve grams).As a red cell I supply oxygen and nutrients.

Each female has two ovaries in the lower region of her body unlike barnacles where the ovaries are found in the head. Baby girls are born with five million pre-eggs of which two hundred to three hundred and seventy five Elsie eggs can develop in a life time. Elsie Egg numbers vary in animals. Cod produce over one million Elsie eggs but only five may hatch. Oysters produce over five million. Elsie eggs at each ovulation and only one may hatch. Roundworms inside humans can produce twenty six million Elsie eggs.

 Chickens can produce an Elsie egg per day while termites produce over two thousand per day. Of the million larvae produced by corals only one may survive. Human Elsie eggs are the largest cells in the body and every human has spent half an hour as a single cell. Sharks have the biggest Elsie eggs excluding those of birds. The biggest diameter of an animal Elsie egg in its shell was thirty centimetres. Pigeons however need to see another Elsie egg before they can produce one and often false Elsie eggs are put in some bird nests to encourage breeding or to stop them producing too many Elsie eggs.

Women have monthly automatic Elsie egg production, producing Elsie eggs every 14th day of their reproductive cycle. Some mothers can feel the production of ova, the finger like projections tickle Olive ovary to encourage release. Elsie egg passes down Fiona fallopian tube helped by other little finger like projections Cecilia. Five days before and five days after ovulation a female can get pregnant.

Precocious puberty (capability of producing Speedo sperm or Elsie eggs) has changed over the years. In some children it is 9-14 in boys and 8-13 in girls. Puberty was eleven years old in grandmother's time, in your mother's time eleven and three quarters and is now normally ten and a quarter years. Obesity, lack of exercise and increased broken homes are said to be resulting in lowering the age of puberty. Some girls have been known to have periods at the age of 2 and develop female's characteristics by 4 years. The effect of obesity is due to high fat levels increasing the amount of the hormone leptin which could trigger puberty. The youngest mother recorded was just under six years old and even a 5 year old child was found with a baby (twenty two weeks old) which initially was thought to be a tumour.

All your reproductive cycles are controlled by hormones. Reptiles do not have sex and hormones and sex is controlled by the temperature of Elsie eggs. Some animals, like buffalo, spray the earth with hormones in their urine. In the male the testosterone hormone in Tony testes (highest at 0600-0800 hours) helps produce and develop Speedo sperm. In the female the follicle stimulating hormone (FSH) from Brian brain starts the whole reproductive cycle off. This hormone (in Pippa pituitary gland) helps the pre egg (follicle) to develop into an Elsie egg, and encourages Elsie egg to produce another hormone oestrogen. FSH is also found in males and the level rises at eleven to thirteen years of age and stimulates Speedo sperm production.

Another hormone from the same part of Brian brain is called the luteinising hormone (LH). This helps release the developed egg into Fiona fallopian tubes and also stimulates male hormones in the male to produce maleness i.e. growth of Lincoln larynx (giving a deeper voice) and development of Harry hair etc. The hormone oestrogen is produced from Elsie egg while it is developing and progesterone (from the cells around Elsie egg). Both affect a change in Wendy womb lining which thickens and more blood vessels (more travel for me as a Fred red cell) are produced so that Elsie egg, if fertilised, can implant and extract food and nutrients. Oestrogen also initiates breast growth and pelvic widening and progesterone causes mood change and temperature increase. Elsie egg only lives a day if it is not fertilised. A fall in the amount of progesterone which occurs if Elsie egg is not fertilised, initiates menstruation (loss of Wendy womb lining). Every twenty eight days (menstruation) one can lose thirty to one hundred and eighty cubic centimetres of blood. Unlucky for me as a Fred red cell, if passing that way. Menstrual cycles can be synchronous for women living together who are affected by the natural pheromones produced.

The hormone androgen helps the sexual drive. Blood flow increases with arousal and breasts can increase in size by twenty five percent. The face also has increased blood flow and appears red. Smells also help relationships, unless they are bad, that is the smells. In humans the female makes herself look more attractive while in animals the reverse is true, just look at a male peacock. External displays, colour, expression, behaviour and protuberances also help with relationships and sexual selection

Men are attracted to rounded contours e.g. breasts buttocks and shoulders. Women it is believed look for clues to check out males. They are more interested in investment if they have children and beauty is not that important. Camels dangle an inflatable sac out of their Moby mouths covered with Sally saliva so that it looks white and attracts females.

The hormone serotonin in both sexes helps develop passionate thoughts while biphenylethylamine (which is also found in chocolate) produces the fall in love bug. . Just remember you lose twenty six calories in a single kiss, good for all weight watchers! Years ago it was said that women if they fixed their Ellie eyes on a man they wanted as a partner then the child born would look like him.

At sexual intercourse the erect Percy penis enters the erect enlarged Valerie vagina. The Valerie vagina is the most absorbent organ in the body. In a hyena Valerie vagina is like a Percy penis so can present difficulties. Labial flaps (major and minor) are found at the entrance to the vagina. In some large sea animals these act as claspers and in some human races these may also be enlarged and hang down to as much as thirteen centimetres. Sex can burn up three hundred and sixty calories per hour but normally doesn't last that long. Sex also encourages production of high rather than low density lipoproteins which cause Happy heart disease and thus can lower your cholesterol level.

It is reckoned in humans there are over one hundred million copulations every day. Birds and chimps can mate over one hundred times in a day. In the UK people on average have sex one hundred and eighteen to one hundred and thirty five times per year. Some queen bees tear their male apart when mating and mantis just eat their male partners. Normally people do not die when having sex but can when they are under great stress or in a traumatic event. You can even get pain relief when making love as endorphins the body's natural pain killers are produced and also a hormone is released which promotes health, immunity, growth and repair. This is very similar to the situation to ensure the female is healthy when pregnant. Single slugs fertilise their own Elsie eggs so need none of this procedure. The female panda is only on heat seventy two hours per year so you really have to work out temperature changes which can indicate receptive times.

In order to get to an Elsie egg, Speedo sperm have a hazardous one metre long journey. Firstly they have to brave an acid bath in the Valerie vagina. The cervix canal is only two and half centimetres long with lots of fibres and tough tissue. Speedo sperm then decide which tube Elsie egg is coming down (wrong choice, then they are doomed). They then fight against the tiny Harry hairs (Cecilia) and fluids that are flowing in the wrong direction carrying Elsie egg down from Olive ovary. However, three thousand sperm can make it to Fiona fallopian tube and these can survive 3-5 days before being eaten by the female's white blood cells. Una uterus with Fiona fallopian tubes is held in place by a butterfly shaped wing of tissue. Elsie eggs send out an attractive chemical. .In all it can take fourteen hours for Speedo sperms to reach Elsie egg.

Once Elsie egg is fertilised it starts dividing as it moves down Fiona fallopian tube to the pear shaped Una uterus. This is seven and a half centimetres long and five centimetres maximum across where it can reach the twenty- two cell stage. Una uterus is deep in Elvis pelvis and partly over the back of Betty bladder so is protected. That is why when Una uterus contains a baby it presses on Betty bladder making women run to the loo.

Elsie egg raids blood for nutrients for growth. Elsie eggs can be incubated in other Wendy wombs of animals such as cattle in a rabbit Ellie egg cell invades the tissues of Wendy womb which grow over it. Elsie egg stage to the size of a baby is a six billion times weight increase. With successful fertilisation as a red cell I am safe as I go with the flow of blood. The ball of cells originating from Elsie egg divide into tissues developed from stem cells. They produce 3 layers of ectoderm (for Slinky skin, Gillie gut and Nellie nerves)/endoderm (organs Lucky lungs and Gillie gut) and mesoderm (Billy bones, Mollie muscles, blood, Katy kidneys etc.).

At twenty eight to thirty five days the pre-embryo has gill slits and a tail rather like a fish. The lower slit develops into the lower jaw. By week four this ball of cells is called

157

an embryo and an umbilical cord starts forming. When pregnant Una uterus expands to twenty times its normal size from a mere 30 grams to one kilogram! This is caused by increased Mollie muscle fibre size and fibre numbers which will fill the whole abdomen. It will never again return to its original size. Stella stomach and Betty bladder are squashed so that they feel full quickly and Betty bladder wants to pass urine faster. The bowels are also squashed so that this can lead to constipation. The spine curves and other organs are also squashed including Lucky lungs. No wonder women can't wait to give birth.

Happy heart pumps more, gets bigger and is pushed to one side. A massive amount of us Fred red cells are produced in order to carry extra oxygen to the baby and as the mother's As Happy heart beats faster we are flung faster around the body. Also around us Fred red cells in the plasma now is a lot of sugar. Some mothers are slightly diabetic at this stage and all will become fatter as well.
.

The baby then floats in the amniotic sac containing amniotic fluid which is released when the water breaks at birth. This is kept at 37 ° centigrade and is changed every three hours. The amniotic sac is anchored by the umbilical cord which ends in Penelope placenta and drains the mother of nutrients and collects waste products. This sac protects the baby, cushioning it from outside. While floating in the amniotic fluid, babies relate to sounds outside Wendy womb such as the mother's voice. Even when born the baby can still sense the mother's mood and will relax when the mother relaxes.

During pregnancy calcium and iron are two very important minerals needed to maintain the health of the woman and also to prevent problems later on in life. Also during pregnancy the blood volume can increase to prevent the mum dying before birth. The woman can gain thirty percent more blood.

Penelope placenta started exploring and invading the mother's tissues when Elsie egg was fertilised due to a hormone. This process triggered morning sickness which is caused by waste toxins produced by the baby. Penelope placenta is eventually like a big red ball, often only one or two cell layers thick around the baby and forms lots of sea anemone like arms (Victor villi) into which we as blood cells flow. Often pools of blood are found where exchange occurs. The one or two layers of cells will act as an exchange surface for food oxygen and waste so it is important to be aware of what the mother eats or breathes. Certain foods, for example some cheeses, fish or unpasteurised milk, are a no go especially where they may trigger allergies or contain harmful Bronwyn bacteria.

At five weeks the embryo is two millimetres long and has a tail. A lot of organs develop at seven weeks when it is just over 1 centimetre long. The digestive system begins to develop as a primitive Gillie gut even earlier than the foetal stage. After 2 months from the start the embryo looks human and has a foetal Spencer spinal cord. By twelve weeks it is nine centimetres long and has the human form of a basic baby. By twenty weeks external features are developing. The early embryo will have two hundred and fifty neurons produced per minute Its Ellie eyes are opened at twenty six weeks and by thirty two to thirty six weeks Lucky lungs have developed and fat is deposited giving it a baby shape. Growth is fastest in the last three months just before birth.

Inside Wendy womb a baby will swallow a cup of embryonic fluid every day. Its senses are blurred as sounds are felt and smells are tasted. The skull sutures are not fixed so that soft Billy bones of the head can pass through

Elvis pelvis at birth. You have stem cells in your bodies which can be turned into Elsie eggs or Speedo sperm, or these cells can be fused with a normal body cell as in cloning.

Each Speedo sperm and Elsie egg is different, not only because you can choose the partner but because chromosomes cross over and exchange genetic material before each Elsie egg or Speedo sperm is formed. Your child is the result of a lucky dip (this is called meiosis). In Speedo sperm the energy producing parts (called mitochondria) which contain some other Diana DNA (genetic material other than that in the nucleus) are destroyed at fertilisation. But in the female Elsie egg these mitochondria survive.

There are some things in nature which are rarer and must be a mistake. A woman usually sheds one Elsie egg from her Olive ovary, but if two leave and are fertilised by different Speedo sperm you get non- identical twins. Sometimes, often due to emotion or stress, a single fertilised Elsie egg can divide and this can produce identical twins (a form of natural cloning as both have the same DNA). However these may not be completely identical mitochondrial Diana DNA can change when the cell splits. Identical twins have the same Penelope placenta while in non-identical twins each have their own Penelope placenta. Also you sometimes can get a boy and a girl not identical as sexes, where the cells have too many sex chromosomes.

Fertility treatment can often give multiple births and for quads this could result in one pair of identical and one pair of non-identical twins. The chances of quads are 1 in seven hundred and twenty nine thousand and chances of identical quads are one in sixty four million. Many conceptions may have involved twins with one of the embryos absorbing the other.

Gestation in humans is around two hundred and sixty seven days, an elephant six hundred and forty days, cats and dogs sixty five days. Often women at this time will start getting more active, showing a primitive nesting urge like birds getting the house ready and tidying up.

Signs of birth include a tightening of Stella stomach and contractions which gets the body ready for birth. There is also dilation of the cervix in readiness. The baby may also 'drop' in Wendy womb reducing pressure on the inside of the mother. Near the onset of birth Valerie vaginal discharge can increase and a lot of fluid is released sometimes with some blood (a show). As soon as the waters break labour often will be in twenty four hours or can still take several weeks.

Normal birth can take 6-25 hours, the first time being the longest. Birth, like death, often peaks between midnight and noon. Some animals such as scorpions can reabsorb their young so they are not born. The lower the metabolic rate for animals the less offspring produced and longer generations.

Birth has to occur when Penelope placenta becomes less efficient due to increasing baby demands. The baby then has to survive a completely different environment. Mothers need to breathe deeper during birth as the baby needs more oxygen and us red cells rush around more. At normal birth the head of a baby is bent 90 °. The size of the baby and shape of the birth canal are two factors that determine if a normal birth is possible. Girls tend to be broader at the hips and boys broader at the shoulders. Evolutionary wise, the size of the canal in Elvis pelvis is decreasing. The baby's head is too big. Even now the skull Billy bones move with soft spots (fontanelles) allowing this to happen at the front, back and at the sides so that its

head is pointed at birth. This is a feature adapted by early
Mexicans where heads were wrapped and remained
pointed. A baby's head normally may look lopsided, but
that soon changes. If the head was any wider then women
would have to crawl to deliver the baby.

Hammerhead shark babies are born head first so have to
have their hammers folded back when born. The decision
to have a caesarean often depends on the size of the baby,
the size of the birth canal and past records of health or
birth problems. There are many different choices when
having birth which include water birth and even using
birthing balls.

The umbilical cord, the link to Penelope placenta for food
and waste transport is an inch wide and has no Nellie
nerves. It is bigger in boys than girls. The length of cord
varies from 0.18-1.12 metres. In boys it is slightly longer. If
birth occurs when standing then it is possible that the baby
could be damaged by hitting the ground. The cord will
break with a weight of over 4.5-6.4 kilograms and can
stretch to 12.5 percent of its length.

At birth the umbilical cord is clamped and cut. It has also
been known for mothers to bite through the umbilical
cord if no scissors are available, which is probably what
happened in the past. The length left behind attached to
the baby drops off due to dry gangrene within 5-10 days.
As a red blood cell in Penelope placenta I could travel 4
miles per hour and could be lost when this is removed.
Penelope placenta resembles a cake and in some countries
after delivery it is dried and eaten. It also can be freeze
dried or made into pate or even gold plated and displayed
on a wall. After birth it takes six to eight weeks for a
woman's organs to get back to some sort of normality.

Even though a new born baby can excrete its own body weight every sixty hours, babies, like small animals, appeal to parents. This is due to their having a large head, large Ellie eyes, round bodies and raised cheeks. This is over - exaggerated in pictures of aliens but having the opposite effect sometimes. The fontanelles (soft spots in head) close up after eighteen to twenty four months so it is important to protect the baby's head at this site up to this time. Babies can be born at just over two weeks, but can survive when removed by caesarean being so small as to fit in the palm of one's hand.

It is estimated that over two hundred and forty babies are born worldwide every minute. If children have babies young then Elvis pelvis may not have developed enough to provide a wide enough birth channel. The record number of children born to a single Russian woman is sixty nine in the 18th century. However, a pair of rats can have over fifteen thousand offspring in a year. Mice can have thirty - four in a single litter. It's a good job humans are not as prolific as it would fill the schools. Babies often suffered in Victorian times and either died and even were killed or given away where mothers could not feed them. The king of Siam had nine thousand wives and died of syphilis. Some monkeys kill their babies so they can mate again which may help regulate population numbers. Some frogs hold tadpoles in their mouths until they develop.

Most women have two Bella breasts but some can, in rare cases, have more. They may also have supernumerary nipples above the breasts or in the armpits like lemurs. Many animals have more nipples but you would have problems with multiple bras. During pregnancy due to hormones and increased blood supply Bella breasts swell up and double in size (big boobs). This can hurt due to blood vessels expanding. Breast feeding can go on for years. Suckling sends a message to Brian brain and the hormone prolactin is released to stimulate the milk producing Gladys glands, In China, breast milk has been used for cooking. Wet nurses, not so common now, provided a continual supply of milk for babies whose parents could not feed them.

 Breast feeding needs to continue for up to four to five months before the baby is weaned onto other foods. Even today in rare cases some mothers are still feeding their children up to 8 year olds, even their husbands are sometimes allowed to have a drink! Puppies and tiger cubs have been breast fed by humans and humans breastfed by animals. The yellow pre breast milk, apart from being a perfect food with high protein levels, provides antibodies against foreign proteins (antigens) in your body and makes you less susceptible to diabetes, heart disease, cancers, allergies and other nasties which you have adapted to over the years.

After three days normal breast milk rich in sugars and fat is supplied as well. The amount of cream present in milk may vary during the day. In cows more cream is present in the morning, less at midday and even less in the evening. Polar bear milk is yellow in colour as it mainly consists of fat. A humpback whale baby drinks five hundred litres of milk per day. Men have the same breasts as females but not the hormones for development. With hormone treatment even men can breast feed babies! Good job it's not widely known. Also it is believed that breast feeding can mobilise the mother's fats so protecting her from Happy heart problems for as long as she feeds.

All babies are born with a downy Harry hair, which is soft and woolly .It usually appears in the fifth month of pregnancy (maybe the remnants of the monkey in you). A similar Harry hair is seen in anorexics. You also keep a downy Harry hair until puberty. Harry hair follicles also become ready a few days before birth. Babies when born are covered in a cheesy varnish of Slinky skin scales and follicle secretions which makes them slippery. Babies, like some other animals, will exhibit a reflex action when dropped in water face down. They will keep their Moby mouths shut and Ellie eyes open and start swimming. This is lost in a few months. This may indicate that you evolved from the sea and fits in with the gills you had as an embryo but not as a baby. New born babies and up to two weeks after birth, babies can have a strong grip and can even suspend themselves for two minutes.

After birth you need to make sure that the baby is wrapped up. This is because the baby's body surface to volume ratio is five times larger than an adult's and it has little fat so it can lose heat quickly especially in colder climates. When asleep their body's metabolism does not increase when they are cold. Babies though do have some brown fat which can generate heat. Probably in ice age conditions back in the past many babies were lost due to cold but saved possibly by wrapping them in the thick fur of mammoths. They may possibly have been a bit more hairy also. Human babies cry when hungry, need company or have Stella stomach pains. Some babies can die if not held. Urine smell may be recognised by a baby.

In humans, unlike some animals, Brian brain is not fully developed when born and needs another year for this. If it were fully developed it would be squashed when passing through the birth canal. A 1 year old can have 30 percent fat in its body. A baby's body weight can triple in the first year. Delay in development, as well as the development of other knowledge through childhood to adult, allows humans to respond and adjust to their environment and needs which is essential for your success today
Humans have live births. Some animals that produce Elsie eggs invest a lot of pre-birth care. Emperor penguin

females deposit their Elsie egg on top of the male penguin's feet while they trek off for miles to replenish themselves with food before returning to feed the hatched chicks. The poor male has to endure temperatures well below -80 ° centigrade while waiting, knowing that any sudden movement could make Elsie egg fall off his feet and freeze the developing chick instantly. The male, although starving keeps inside its Gillie gut as a last resort a food reserve to give the chick just before the mother returns. The poor male, very weak, then has to walk many miles for weeks to look for food for itself.

The fastest growth for a child is in the first year. It takes up to a year for children to gain normal vision and good hand, Ellie eye co-ordination. This is position sense so they can tell where their arms and legs are without looking. Girls tend to put on most weight between ten to fourteen while boys are later twelve to sixteen. In the first eight years the immune system develops. Children can be brought up by wolves; dogs even ostriches and survive as wild humans.

Secondary sexual characteristics occur due to hormone production at puberty.
Testosterone causes a deep voice in the male (due to growth of vocal chords). This hormone also results in growth of Charlie chest armpit and groin Harry hair, and Percy penis enlarges, Tony testes become functional, spots are found, and Mollie muscle and Billy bone size increases. Oestrogen hormone in the female results in the enlargement of Bella breasts, nipples, pubic and underarm Harry hair. Also this hormone causes increased fat deposits under Slinky skin, and the onset of menstruation. In some cases women may have beards and some tribes are very hairy all over including faces.

Reproductive system problems

Many superstitions exist to help women to become pregnant-eskimo women placed dolls under their pillow. Nature favours women who are able to produce children, after this period nature does not seem interested.

A serious complication in pregnancy is pre-eclampsia where there is a defect in Penelope placenta so that blood pressure increases and can affect Katy kidneys. With high blood pressure or other problems a baby may be induced. This involves the use of a hormone as a tablet or gel released as a slow release pessary placed in the Valerie vagina. This should take a day to work after which if not successful you may have another dose. Hormones can also be introduced through a drip. Also forceps or suction may be applied more often after induction.

Failing this procedure a caesarean can be given. A 'C' section is under local anesthetic freezing the area involved. A horizontal incision is made in the abdomen wall (a Bikini Cut) just over the pubic Billy bone avoiding Stella stomach and Mollie muscles. An incision similar to the previous one is made in Una uterus, the amniotic fluid is sucked out and the baby delivered. The umbilical cord is cut and lastly Penelope placenta will be pulled out. The Mollie muscle layers etc. are then sewed up. Babies as small as fifteen centimetres have been born in this way and still survive. Some though may have defects and require microsurgery.

Ectopic pregnancy is where the embryo develops in Fiona fallopian tube or falls out of Fiona fallopian tube. It can attach to organs, even Lily liver as it has a good source of blood and can remain there for many years.

167

Women over forty have more chance of underweight babies and other problems may occur. One can tell the sex of a baby after 6 weeks, which can in some cultures encourage abortions where male children are more advantageous and families cannot afford the dowries for their daughters.

After birth oestrogen drops a lot and a chemical imbalance occurs which can cause postnatal depression which can last indefinitely. Previously before and during birth it gave the mother a high as it increased the happy hormone. Women with small Elvis pelvises or premature babies and other problems such as no success with induction may need to have a caesarean; I lose a lot of my red blood cell friends this way. Some women just prefer this non-natural method of delivery.

Sex hormones can affect cancer. Some forms of the female sex hormone oestrogen encourages breast cancer but delays Patrick prostate gland cancer. Phthalates and other chemicals in water can mimic female hormones. Often male fish become female as do polar bears.

Fathers also do not get off lightly. They also have Gladys milk glands but these don't develop. Some men develop their own symptoms while their partner is pregnant (sympathetic pregnancies). These can include severe Stella stomach pains, spasms, and even weight gain. There may be hormone changes in the male at $3^{rd}/4^{th}$ months and near birth of their partner's pregnancy.

One in eight of all people could have started off with a twin which may have been reabsorbed as pregnancy progressed (Vanishing twins). Some inner twins may feed on the other inside. It has been suggested if you are left handed you are a mirror image of another twin reabsorbed before birth Mirror image twins can exist with their organs on opposite sides. Sometimes the remains of twins can be found inside babies or even adults at any age. Adults can have cysts removed which contain teeth and Harry hair. In one case a 0.46 metre foetus continued to live off a 7 year old child as a parasite before it was found. Also in some cases one can find conjoined twins where twins are attached to one another. This becomes more of a problem as they become older especially where major blood vessels or Nellie nerves are involved. There are famous cases where twins remained joined together as adults. Most are joined at Charlie chest. Siamese twins who were joined at Charlie chest actually fathered twenty two children. The Hilton sisters were joined at their sides but were not allowed to marry under Americas laws. They made a film about themselves which changed the attitude of Americans towards freaks.

 Other examples exist where parts of a parasitic twin remaining include an extra leg, half a body, extra genitals and even a small functional head on the top of a fully formed adult. Cows can be found with eight legs and two tails Siamese twins suggest that some similar parts fuse together and don't release. Sometimes some parts of the internal body can be found outside the baby's body at birth. In many cases this can be rectified. Happy heart in one case was found in a baby's hand. There also may be deficiencies inside - some girls are even born with half a Wendy womb but still can have children. Some women are born without a womb but now womb transplants have proved successful.

The body is very resilient and can adapt to losses of body parts which are the result of genetic malformations or sometimes drugs. Some of these people become very mobile and exhibit talents which would not have been expressed if they had been like ordinary people. In some cases a child's gender may not be certain at birth and the genitals are aligned to one sex or the other by surgeons. It has been known for babies to be assigned as a male and only at puberty when menstruation occurs through Rosie rectum is this mistake realised.

Fertility rates and Speedo sperm counts in males especially in Europe are falling. The western human male now only produces half the number of Speedo sperm as his father did and Speedo sperm is less active. Some women have problems in conceiving as they are allergic to their husband's Speedo sperm. This means Speedo sperm is considered a foreign protein and will be attacked by antibodies when it enters the body. Foetal tissue can also be recognised as a foreign body and can be attacked by white blood cells so again there is a need to suppress the immune response.

Winston white cells already break down any unwanted Speedo sperm. A similar thing is where one partner has rhesus negative blood and the other rhesus positive and the baby's blood does not match the mother's. A bleed can initiate baby destruction. There are many hormone deficiencies which can affect the fertility of the man and woman, as well as problems with implantation. Mobile phones could, it is believed affect fertility in a man due to the electromagnetic fields. This may also be due to stress from having a phone.

In each step of Elsie egg development something can go wrong, especially in women near menopause. Miscarriage is believed to be in some cases nature removing genetically abnormal progeny ensuring maximisation of future survival. Some babies are born with the amniotic sac and fluid still around them so no waters are broken. Also there may be a problem where Elsie eggs and Speedo sperm are cloned from a person's own cells.
Some women just can't get pregnant. Many methods now exist to help fertilisation and pregnancy. Speedo sperm can be frozen for 5 years. Methods include artificial

insemination where Speedo sperm is introduced by a syringe, where Elsie eggs are fertilised in test tubes in In-vitro fertilisation and the fertilised eggs are replaced into Wendy womb with added hormones to help implantation. In vivo fertilisation involves planting Elsie eggs and Speedo sperm inside Fiona fallopian tube. However some women have had perfect babies (with fertility treatment) when aged over 60 years. There still can be complications and also side effects from the drugs given. This, unfortunately, throws up questions of childcare etc. by the aged parents later.

Often hormones such as FSH are given to the woman to increase fertility and produce many Elsie eggs which then can give rise to multiple births. Sextuplets have been born in six minutes as a result of using a fertility drug, but with large numbers normally they don't all survive as blood flow to uterus is affected or they are affected physically in some way, but things are improving.

With cancer, where Elsie eggs would be affected genetically if not destroyed by the radiotherapy, a layer is removed from the patient's Olive ovary. This is cut into sections and frozen in liquid nitrogen. After treatment this tissue is thawed out and transplanted next to Olive ovary at the end of Fiona fallopian tube so that good Elsie eggs can develop. Those people working in dangerous jobs often also have their Elsie eggs or Speedo sperm frozen in case of accidents so that they can still have children even if a disaster happens. There is around thirty percent chance of getting pregnant with two Elsie eggs after they have been frozen but nobody knows what turns an embryo into a baby.

In the future cloning may be a way of producing children. Cloning for identical copies has been done in many animals in some cases to get identical copies of a pet that has died. Cloning may produce an identical copy of an animal or human but how they develop and what they develop into is still governed by the environment at that time. It involves taking any cell from a normal person's tissue from which you want to clone. Its nucleus in the Elsie cell is then removed and injected into an Elsie egg cell from a donor whose nucleus has been removed. The cell is then given a zap of electricity which starts it dividing and then it is placed into a donor or the original Wendy womb to develop. Cloned animals are not always successful; some have problems with increased ageing and other disorders. The cytoplasm from the donor cell may also influence the final product causing disasters which one would not expect. Mitochondria in the female egg cell can be faulty resulting in an ill child. Mitochondrial replacement can be carried out from a donor but it is believed that the traits of the donor will be included together with inherited traits from both parents so this may result in a three parent baby which may cause ethical problems.

This Diana DNA has not changed drastically from the one that was in the first woman (mitochondrial Eve). This has passed down through the generations to you. Wow. Human cloning has reached several cell stages of development and is useful for producing stem cells but there still remains controversy over producing humans in this way. Stem cell introduction into a body if from a different sex, can mean a Y chromosome may get into a females body. Elsie egg selection is possible in order to help a member in a family who would benefit from their brother or sister's compatibility.

Ageing and reproductive problems
After all Elsie eggs have been produced in older women, they often have hysterectomies. This involves removal of Una uterus and cervix or Fiona fallopian tube is closed.

The tissue holding Wendy womb in place can often slacken so that Wendy womb hangs out of the Valerie vagina. This is prolapse and can also be caused by lifting heavy weights. Wendy womb has to be pushed back in place or held in place by a metal mesh. Nature has not produced a device for removing this organ once its use is over. This can also be carried out earlier if there are serious fibroids which cause pain and bleeding .After thirty five fertility falls and the lining of Wendy womb becomes thinner. Menopause in women occurs in their fifties usually but it has been known in teenagers and also to continue up to ninety years old. Oestrogen production is reduced as Elsie egg development ceases. Some believe you can eat plants that have chemicals that act like oestrogen e.g. soya beans, lentils and chickpeas. Women with the menopause can experiences hot flushes (red face), sleep disturbance, mood swings, depression and other symptoms.

There is said to be a male menopause where testosterone levels fall also causing night sweats. Even men over one hundred years old can father a child. Some say as men age and testosterone male hormone levels fall they tend to become more women like and more caring.
 After menopause in women oestrogen levels fall and as said before this can lead to calcium loss so hormone replacement therapy is used. They may also become more hairy as the female hormone is reduced and they may also grow beards and may even become male like.
A lot of men over fifty especially with Happy heart problems or diabetes can become impotent. Viagra and other drugs are now commonly used to help with this.

12 DANGER KEEP OUT-WAR IN THE BODY (BRONWYN BACTERIA AND FRIENDS ARE ABOUT)

Some Bronwyn bacteria have adapted many years ago to life on and inside man and some believe you originated from them. Ten percent of your whole body weight is made up of micro-organisms. Wow, perhaps you could lose a few and lose weight. As you move you need a repair system to survive, but while repairing you need to defend yourselves against micro-organisms all around you and especially on your bodies. You are becoming too clean today, not exposed to as many germs as in the past so are less resistant. Secretions which are intestinal, respiratory, from urogenital tracts contain antibodies as do those from Slinky skin, Ewan ears, tears and breast milk. Disinfectants or materials containing antimicrobials used in the house and elsewhere tend to select for certain Bronwyn bacteria. These are able to survive and become resistant and multiply.

Where hosts of people are close together there is increased transmission. As has been seen already, my journeys around the body can vary in speed, and my environment is continually changing. Water content, salt, nutrients, hormones and waste products, plus many other things have entered your blood from Gillie gut. As well as this form of entry, bugs can enter openings in the body such as Moby mouth or Noddy nose leading to the respiratory tract. Other opening include the urogenital tract, broken Slinky skin or from bites from animals.

Insects are great vectors of disease. The housefly sucks up semifluid material from meats and faeces and regurgitates this with enzymes to take in the meal. Bronwyn bacteria etc. can be in the fluid and also on the insect's feet.

Another threat to me as a red blood cell and especially to my neighbours, the white blood cells, are Felicity fungi, Veronica viruses and the Basil Bronwyn bacteria family, There are as well as other foreign proteins which could set up an allergic reaction. They can get through every orifice in the body and are especially dangerous when they get into blood vessels next to me .e.g. by a cut. In the 1830's they had thirty seconds to amputate and close up a wound before Bronwyn bacteria got in. Bandages help prevent infection. The barber's pole seen sometimes outside hairdressers represented red for blood and white for bandage. Barbers were barber-surgeons often called on for minor medical procedures e.g. tooth pulling, setting fractures, lancing boils and especially annual bloodletting. Bronwyn bacteria are everywhere.

One drop of liquid can hold fifty million Bronwyn bacteria! Swimming pools are a good source of bacteria as well as areas of the coast near sewage outlets. Bronwyn bacteria already exist over the body reaching as many as thirty two million per square inch or ten million per square centimetres on Slinky skin. Your body can hold over two hundred and thirty grams of Bronwyn bacteria. Mobile phones can have more Bronwyn bacteria than on a toilet seat. The worst place for Bronwyn bacteria are under your Nelly finger nails. Pull your tongue out and that white layer can be all Bronwyn bacteria. Your tongue holds fifty percent Bronwyn bacteria in your Moby mouth. Seventy five percent of bad breath is due to these.

There are over one hundred million in your Moby mouth. Sugars stick on teeth which with the help of bacteria can produce lactic acid and dissolve the enamel. At the back of your Moby mouth is a ring of lymph nodules which form a ring of tonsil tissue which protects the throat.

Bites from animals have Bronwyn bacteria on teeth and are often dangerous e.g. Rabies. Komodo dragons have a filthy Moby mouth and have a bite that can kill you. In defence your body has produced secretions, enzymes and acidity to protect the outside but these are not always sufficient.

Most Bronwyn bacteria on your body are friendly Bronwyn bacteria which help fight off the others. We have built up defences against as you evolved and also gain these from birth through breast feeding. Howard Hughes, a successful Hollywood director, had OCD and collected his faeces and urine. He avoided germs, by always washing his hands. He kept his curtains closed, placed tissue on the floor for people to walk on and did not cut his nails. Toshers in Victorian time sifted excrement for jewellery in the sewers and although stinky, survived disease.

Mud larks on the banks of the polluted Thames in London also looked for valuable items. HG Wells War of the Worlds illustrated how humans had built up resistance and showed how Martians arriving on your planet were not adapted and died because of Bronwyn bacteria. Unfortunately you may all have specific Bronwyn bacteria. This is seen in your early discoveries of other countries or new tribes where you managed to infect entire populations with micro-organisms they had never met before. In some cases these Bronwyn bacteria annihilated them.

In 1350 nearly half the population of England died out due to the Black Death, caused by Bronwyn bacteria carried by black rats. Lymph nodes became swollen in arms and legs (buboes) where the body could no longer defend itself. These buboes turned purple to black and Bronwyn bacteria invaded the body so that tissues liquefied and rotted. Houses were quarantined and the call 'bring out your dead' was used to remove bodies from houses to plague pits.

Tuberculosis of Billy bones and Lucky lungs was also a killer, and is still today in some countries. Many of those infected wee sent to seaside locations and often placed in chairs or beds on balconies so as to help with heir breathing. Some micro-organisms you have never been able to become resistant to, but you give ourselves injections of parts of them so as to become resistant. Bronwyn bacteria also live in Gillie gut.

Greater than ten million are found in your bowel where they entered with food or drink. Babies have no Bronwyn bacteria in Gillie gut so need to avoid non-friendly ones entering. Some Bronwyn bacteria produce vital substances such as vitamin K in Gillie gut. Babies do not have these Bronwyn bacteria at birth so an injection of vitamin K has to be given. Disposable nappies have reduced the transfer of Bronwyn bacteria from baby's poo when washing nappies to other washing.

Once inside the body Bronwyn bacteria compete for food and minerals and can attack and destroy cells. Depending where they are, they can have serious effects on metabolism and produce toxins. Once Bronwyn bacteria are detected inside the body blood flow is increased as well as body temperature (fever) to shift and cook them. At the same time the immune system (antibodies with an army of Winston white cells) tries to deal with the crisis.

Pus holds Bronwyn bacteria and can sometimes not be yellow but blue due to a different kind of Bronwyn bacteria which produces a chemical called pyocyanin. The Lesley lymphatic system produces Winston white cells for defence. Removal of Lesley lymphatic glands or Spiro spleen which also produce defence cells means your body is protected less and you may be vulnerable to infection. If you die all body defence is lost and Bronwyn bacteria will multiply and will help digest your Gillie gut and organs. Once a specific Bronwyn bacterium is dealt with, the body has built up a supply of antibodies to deal with these Bronwyn bacteria again if they revisit. Bronwyn bacteria in the blood can produce toxins which affect Nellie nerves or organs. Septicaemia, pneumonia and toxic shock can result and are the most dangerous. Bronwyn bacteria are identified by their shape. Rod shaped ones are called Bacilli. An example of rod shaped is E-coli which live in your Iris intestines. Some strains (mutations) can really be harmful especially if they escape from an ulcer or burst April appendix. This can cause septicaemia or blood poisoning.

Gas gangrene is due to Clostridium bacteria multiplying, especially in deep wounds and producing a bubbling gas in rotting tissue. The only way to stop the spread was to cut out the infected area or amputate infected limbs. Legionnaire's disease is also another common rod shaped Bronwyn bacterial disease causing lung infection. Bronwyn bacteria thrive in water showers and air conditioners. As you get older you lose a lot of friendly Bronwyn bacteria, so probiotic yoghurts are useful to top up numbers. Bronwyn bacteria in your Ewan ears are captured by the wax so don't be too keen to remove it all. There is more wax in peoples Ewan ears in the city than in the countryside.

Cattle have E-coli in their gut and can cause salmonella, so food hygiene must be adopted when dealing with meat. Care has to be taken therefore in eating undercooked meat. You excrete every day one trillion Bronwyn bacteria. .Hygiene is important especially where animal products exist next to other foodstuffs. The sell by date on foods is not for fun ignoring this can kill you.

Round shaped Bronwyn bacteria are named Cocci. An example is <u>Streptococcus</u> which can cause a sore throat. <u>Staphylococcus aureus</u> another Coccus is present all around you especially on Slinky skin and up Noddy nose. It is common in pus filled spots on Slinky skin. Antibiotics such as meticillin may not be very effective as mutants continually occur .This has led to many infections in hospital before and after treatment, an example being MRSA(meticillin-resistant <u>Staphylococcus aureus</u>.). In twenty four hours one bacterium can produce seventeen million others. Toxins produced can lead to infections and if found in Lucky lungs result in build-up of fluid (pneumonia) and if in the blood, blood poisoning. Bronwyn bacteria kill Winston white cells and there is a lowered immunity.

One of the worst toxins, deadly in minute amounts is that produced from Botulism. Toxins cause paralysis which can shut down the whole body by switching off the Nellie nerves. However the toxin from Botulism if injected in safe doses, is used today to remove lines from your face in cosmetic surgery

There are also spiral shaped Bronwyn bacteria. These are corkscrew like movers, an example being one that causes syphilis. This was found in Christopher Columbus, Henry V111 and Napoleon. Mercury was used as a treatment for syphilis with deadly effects. The discovery of fungal penicillin by mistake, was a godsend in the war. Alexander Flemming left a bacterial culture in his laboratory with the lid off. Mould from the walls of his laboratory landed on the culture and killed some of them. The mould contained penicillin which prevents Bronwyn bacteria from dividing. Now however   many Bronwyn bacteria are resistant to this. Any Bronwyn bacteria can mutate and there is a continual fight to find new methods to control them.

The problem with all Bronwyn bacteria and other .micro-organisms is their rapid multiplication rate which ensures that at least one new bacterium will be immune to an antibiotic and continue to survive and breed. A few years ago antibacterial kitchen ware and boards were produced but fears that these would concentrate the more resistant virulent Bronwyn bacteria became commonplace. This is similar to the overuse of antibiotics and disinfectants today.

Bronwyn bacteria are essential for decomposition and linked to your mineral cycles. Water helps decomposition. Three thousand bodies are thrown into the river Ganges in India each year and decompose faster than in the soil. Your decomposition in the soil is slowed down by the preservatives you have eaten. High grade honey (manuka) and sugars can kill Bronwyn bacteria in wounds as can prevention by using medical maggots to eat dead Slinky skin around the wounds.

If you thought Bronwyn bacteria were bad, Veronica viruses are a thousand times worse. In 1918 a flu epidemic killed over twenty one million people. They can float alongside me and other blood cells in the plasma and can attach themselves to the outside of cells. They then inject their Diana DNA or RNA into the cells which organise them to produce more Veronica viruses. The cell becomes filled with virus so many Veronica viruses that it bursts and the new Veronica viruses search for new cells. Some Veronica viruses alter Diana DNA of the host cell they invaded and wait for it to divide naturally so that it remains in the new cell.
Veronica Viruses are thought to be more primitive than Bronwyn bacteria as they can remain as infective crystals.

A dry hanky can contain crystals of Veronica virus from a sneeze from a cold, so taking it out from your pocket can release these infective crystals everywhere. A cough can make Veronica virus particles travel ninety six kilometres per hour, while sneeze particles travel at one hundred and ten to one hundred and sixty miles per hour so watch out. Turn away quick.

To destroy Veronica viruses the body needs to build up its immune system producing antibodies against them. Low or inactive doses of Veronica virus are often given as injections to provide the body with resistance to attack. Bird flu comes as a different form each year with different spikes on the outside of the viral ball. You make a guess as to which is the closest and give old people and young children protection by injecting them with an inactive dose. Some Veronica viruses may still be active, the polio virus given to babies can be excreted onto nappies and grandparents were always warned to be careful when changing nappies as they may be contaminated. Warts are caused by a veronica virus which can spread and is often reoccurring when your resistance is low.

Microbes, once they get into my plasma, can target one organ or many.
The body can build up a good immune system. Interferon is produced by most cells and can stop viruses multiplying but not HIV and Aids. The AIDS virus attacks Lymphocytes the large white blood cells which normally fight infection. A thing called a retrovirus invades helper T Winston white cells so help in defence is lost. It can take fourteen years to get full blown as AIDS. All your body defences, which you gained from birth and built up through your life, are threatened. Then other viral, bacterial, cancers and fungal infections can all gain hold. Treatment is by combination drugs, antiviral drugs to stop replication and the slow the progress so your body has

some time to fight. Some people are already resistant and the search for a vaccine continues. Pneumonia as with many critical disorders is often the final cause of death. Ebolavirus is another serious killer. It originated from monkeys bats and other animals. It is fast acting and is transmitted through contact with blood or other body fluids. Veronica virus causes initial flu like symptoms, which leads to vomiting and diarrhoea. This is followed by breakdown of tissue resulting in internal and external bleeding, your blood vessels leaking and collapsing. Wounds do not heal and blood leaves every orifice and blood pressure drops.

Felicity fungi are not such a problem as Bronwyn bacteria or Veronica viruses. They still can extract nutrients from cells as well as block passage ways in Lucky lungs and cause other infections. Some Felicity fungi have enzymes that break down keratin in Slinky skin, Nelly nails and Harry hair. Examples include ringworm, athletes foot and jock itch, which can be transmitted easily. Ringworm lives in damp areas on Slinky skin and under the Nelly nails. Fungi thrive in warm damp areas.

Some Felicity fungi are opportunistic meaning they rush in when a person is weakened by an illness e.g. diabetes (sugar in urine) or AIDS. Diabetes can lead to thrush is often seen in Moby mouth or genitals. Thrush thrives in warm damp genital areas, Felicity fungi can start up allergic reactions and feed on dying cells. Spongy pillows are the worst as Felicity fungi as well as dust mite and faeces, can live in the holes. A pillow can contain more than a million fungal spores of sixteen different Felicity fungi. Verruccas are picked up from other people often in swimming pools. The blood supply is in the middle of the verruca,

Protozoa are single cell organisms which normally live in water and soil. Sometimes, like the one celled protozoan amoebae they can sit in Gillie gut and inflame the abdominal wall lining and can form cysts in organs such as Lily liver. This affects water movement in Gillie gut causing dehydration often through diarrhoea. They eat red blood cells so I am not happy about that and once in the body may travel around in the blood. Some amoebae can even enter Brian brain.

In some cases protozoa have flagella, (many Harry hairs for movement) and are called Flagellates. Giardia is an example and this can produce Stella stomach cramps even on the thought of or the sight of food and send you rushing to the loo.

Malaria, sleeping sickness and toxoplasmosis are examples of other diseases also caused by protozoa which have adapted to living inside humans with other hosts for transmission. All have enzymes for dissolving cells so as to penetrate tissues. In malaria the plasmodium stage enters red blood cells where it multiplies. This makes you anaemic but also infected Fred red cells become sticky and block blood vessels causing chaos here and Brian brain clots. Mosquitoes spread the disease by feeding on infected humans. They are more attracted to people eating bananas so don't eat bananas in bed without a mosquito net. In cattle Bronwyn bacteria make proteins from the grass eaten using protozoa and urea.

Worms such as Felix flatworms and tapeworms also can pose a danger. They are often flattened with little Brian brains and no ways of moving. Iris intestines may be packed with worms and often like tapeworms have large well adapted hooks for attachment. They also have large Moby mouths to gulp in food. The hooks destroy cells and organs and also produce toxins which upset the body's balance (homeostasis). All worms produce large numbers of Elsie eggs in the hope of contacting a new host. Pin worms or thread worms become more active at night and crawl out of your Andy anus to lay Elsie eggs. This can make you feel itchy. These Elsie eggs can get on clothes or Slinky skin and can be swallowed. Worms can produce twenty thousand Elsie eggs and some like tapeworms form cysts in uncooked meat which when consumed can form cysts in organs and Mollie muscles. A five inch one was found in Brian brain. Hookworms can enter the body through the feet and go into Lucky lungs then be coughed up and swallowed and they enter Gillie gut. Cats and dogs can deposit toxicaria worm Elsie eggs in their poo. These can be transmitted to humans and the worm inside the human can travel to Brian brain or Ellie eye causing damage. It is therefore important to deworm cats and dogs especially where you have young children.

Roundworms can cause Elephantitis, which can produce swollen limbs the size of an elephant which block lymph vessels causing the swelling. The Guinea worm larvae lives in water and is eaten by water fleas which are swallowed by humans in drinking water. The worm can come to the surface of the body from the intestine as a burning blister on Slinky skin and may be one metre long. It can be wound out daily using a stick but if it breaks and dies in your body can cause toxic shock.

Ticks, mites, fleas, lice and bed bugs are other examples of transferring infections to humans. They often have an animal host from which they are transferred. They pick up Bronwyn bacteria, Felicity fungi and protozoa when feeding which they readily transmit to humans. They also can cause allergies and irritation. Scabies itch mites burrow into Slinky skin and lay Elsie eggs in the creases and sometimes transmit serious diseases such as plague and typhus. Some mites live in Harry hair follicles and in oily Gladys glands in Slinky skin. One can get one hundred dust mites in a gram of dust. These will eat dead Slinky skin scales. There may be as many as six million dust mites in a bed. Bed bugs hide in very small crevices behind wall paper in the day and emerge at night to feast. More of these are being spread around by travel in suitcases. They also give out a characteristic smell. Some animals such as hedgehogs rely on the presence of fleas to stimulate them and can actually die without them.

Pubic lice attach to Harry hairs in the groin or other areas but don't like Harry hairs on the head as they are too far apart. They make you itch more at night. You can also get crab lice which hang onto pubic Harry hair with their pinchers, and also cause itching. Head lice cling onto Harry hairs on your head and become the colour of your Harry hair. Pharaoh's ants can cause infections especially in hospitals where they can crawl under bandages and move from one patient to another spreading Felicity fungi and Bronwyn bacteria.

Long periods lying down or being bedridden for a time can pool body fluids in places in the body and encourage infections. When epidemics occur not every one dies as some Jenny genes can protect people. This is what is called natural resistance which will be passed down from these people to future generations.
Leeches as a parasite can enter any orifice in your body to suck blood. One was found inside a woman's nostril and had entered while swimming. They are useful in restablishing blood flow where fingers or toes have been re-joined after amputation. They help reduce bruising for facial procedures.

13 DODGY JENNY GENES AFFECTS MY LIFE STYLE

Jenny genes occur in all cells except myself (red blood cell). Jenny gene is a length on the strand of a piece of the tightly bound Diana DNA which forms the X shaped chromosome. Diana DNA has several thousand base pairs. These make up your genetic code. In 2003 they found that human Diana DNA has 2900 million base pairs with thirty to forty thousand Jenny genes. Ninety percent of your Jenny genes are the same as in yeast and you share fifty percent of your genetic code with a banana. Chimpanzees have 98.5 percent of your Diana DNA. This means they are very like you, but the 1.5 percent makes a lot of difference. You have lots of advantages but chimps don't get AIDS. You still cannot recognise small Jenny genes or Jenny genes inside other Jenny genes. One to two percent of your Diana DNA actually codes and the rest is not used? Jenny genes can be switched on for different times for different functions. One an example is to form the longer neck in a giraffe.

Normally humans have forty six chromosomes in a cell, twenty two pairs of normal ones and one pair of sex ones. The female chromosomes are denoted as XX and the male XY in the sex cells. The adders tongue fern has the most chromosomes which are seven hundred and twenty pairs. In sex cells

Elsie eggs and Speedo sperm each has half the number of chromosomes i.e. twenty three. When Elsie egg fuses with Speedo sperm this brings the total number back up to 46. You would expect 50/50 chance of each sex crossing XX with XY but there is usually more chance of boy. This is possible because Speedo sperm with Y chromosome is faster in reaching Elsie egg. Other events however can also lower the chances of having a boy. It was thought one of the female's X chromosomes was inactive but now it seems over half are active and one X chromosome can compensate for genetic faults on the chromosomes. You have dominant and recessive alleles which are expressions of a particular characteristic. A dominant characteristic can override a recessive one. An example of this is red/green colour blindness which is due to a recessive allele (h) which can be dominated by the dominant allele (H) so that the person is not colour blind.

When a normal male (XHY) male is mated with a normal female (XHXh) the results are shown in italics the possibilities here are a normal girl *XHXH* and a carrier girl *XHXh* not colour blind. For male children you could have a normal boy *XHY* or a colour blind boy *XhY*.

Father Mother	XH	Y
XH	*XHXH female* *normal*	*XHY male* *normal*
Xh	*XHXh female* *normal but carrier*	*XhY male* *colour blind*

Jenny genes control what you look like and how you work. Jenny genes can work together and are switched on or off. They often produce proteins for organisation. These proteins are enzymes or hormones which actually do the work. Harry hair colour is genetically determined and affects the amount of pigment present in Harry hair. Several Jenny genes determine the amounts. A dominant brown allele exists which overrides the blonde allele. A non-ginger allele can be present which overrides the recessive red Harry hair allele and stops the red pigment pheomelanin being produced. Red and brown Harry hair types contain carotene. Red Harry hair was found in many famous people such as Winston Churchill, Elizabeth 1, and Napoleon. Blondes tend to have thinner Harry hair than red or black haired people. Some colours may also be similar on Slinky skin. The tiger's stripes are not only in the fur but also on Slinky skin.

Dogs and cats are all examples of animals bred from a single species. Many of these through breeding are deformed either physically or physiologically. The bulldog for example, with its flat face, can have breathing problems and can overheat if stressed.
The main blood types in humans are A and B which are both dominant to the type O which is recessive. Charlie Chaplin was accused of fathering a child. Blood groups were used to see if he was the father. Charlie had blood group double recessive i.e. OO. The mother had blood group A meaning she was either AA or AO. The baby had blood group B which meant it was either BB or BO.

Crossing Charlie's alleles with the mother in no way could produce a B as neither Charlie nor mother had a B but the father did. Charlie could not be the father and the B allele must have come from the father. Unfortunately in 1943 blood groups were not taken as evidence in California, so Charlie was still accused of being the father.

Faulty chromosomes and Jenny genes provide the source for diseases and disorders and many hundreds can now be identified. Genetic screening can be carried out after in vitro fertilisation (in petri dishes) and in some cases the best embryos can be chosen or those closely matching if a sibling needs a transplant. This brings in ethical issues arising from a loss of a genetic lottery, to actually choosing which child you want. Some Jenny genes are useful, some animals that over winter may have mechanisms involved to protect them from the cold. There is a pain Jenny gene which blocks up the sodium channels for Nellie nerve impulses and this is found in those people who can walk on hot coals.

Genetic problems

Mutations can be useful, harmful, life threatening or just a nuisance. You have grown taller your Neanderthal ancestors were smaller people. Also when you donned shoes you no longer found a use for the gripping action with your little toe. Ann Boleyn had 6 fingers and three nipples and Marilyn Monroe 6 toes, a pain when buying gloves and shoes. Some people can have fourteen fingers or fifteen toes; they can even have two big toes due to mutated Jenny genes. Stalin had webbed toes which is said to be useful for swimmers. Lobster people have hands with only a thumb/toe and two digits present.

Some people have missing organs or organs back to front or on the wrong sides or have extra body parts. Cyclopia is where babies have one Ellie eye, Ellie eye orbits not separating, which can be due to a sequencing error on one gene. Holoproencaphaly is where parts of the face are missing and the baby may have one large Ellie eye and its Ewan ears on its chin. A Chinese man had an extra Ellie eye on the left but could not see with it.

Gene defects can affect hormones and enzymes which are important in development. In the past children who had abnormal birth defects were thought a disgrace or even a curse. They were left outside to die or steal and beg. Some were sold off to travelling circuses or shows. Some in groups made up their own ethics for survival.

A common example is dwarfism .General Tom Thumb was only 63.5centimetres tall .He was a great favourite of Queen Victoria and became very popular.

Other examples of birth defects include the caterpillar man or human torso. He had tetra Amelia syndrome and had no arms or legs but was able to move using his body. He could also use his Moby mouth for washing and lighting up a cigarette. The bearded lady also was common in shows and developed a beard 5centimetres long from the age of eight. Microcephalics or 'pinheads' as they were known had tapered heads with large jaws. In Pakistan these are called 'rat children' believed to be messengers of God and are sold for begging in some cases.

Virchow-Seckel Syndrome is an example of a skeletal disorder giving rise to a person that looks like a bird. Here the head is small, with few teeth. The face looks very much like a bird with a long Noddy nose that looks like a beak. Ellie eyes are also large, the jaw recedes and they are often blind. Some people may be born with only the top half of their body but these still are very active and can move fast using their hands.

Mutation can cause the death of cells, or over production of chemicals. An example of the latter is the extra chromosome at chromosome 21 which produces Downs. This is where one chromosome may not move, or gets stuck to another during cell division. This causes over production of proteins and enzymes which increases the sizes of organs e.g. Happy heart, tongue and hands. People with Downs can still live to over sixty five years of age.

In sickle cell anaemia us Fred red cells become crescent moon shaped. This is not a lot of good when we want to collect oxygen from Lucky lungs. Also we become sticky and can block up blood vessels. This is really bad news, especially when you get so tired (as less oxygen). However on the good side, in Africa and the tropics malarial parasites can't get into and survive in us Fred red cells if we are sickle shaped, so there is less chance of developing malaria.

Another example of mutation is haemophilia. This is due to the lack of clotting agent so you become bleeders, losing us Fred red cells at the slightest knock. Queen Victoria's family was afflicted by haemophilia. She was a carrier but was unaffected.

Genetic testing for over three hundred diseases and disorders can be carried out on an eight to twelve weeks Penelope placenta. At fourteen to eighteen weeks after conception the amniotic fluid can be tested (using ultrasound to guide the needle).

. If males get extra X chromosomes i.e. XXY or XXXY this is Klinefelters syndrome. This increase in X means a lowering of testosterone so that breasts develop and they become more female. The male reproductive organs are under developed and Speedo sperm count reduced. XYY can cause Jacobs syndrome where males can have excess hormones is reduced fertility when older and may become stronger and more aggressive. They can become criminals also. A male with YO dies. One can also get super females being instead of XX, XXX or even XXXXX. This affects fertility and intelligence. XO i.e. loss of an X or part of an X in females can cause Turners syndrome, Here the female individual can be infertile, have degenerate ovaries and lower amounts of female hormones.

An example where genetic screening is used for a mutation is for Cystic fibrosis
Here Chromosome 7 is affected. This mutation affects ion flow so less water is in the respiratory tract so there is

thicker mucous. This builds up in Lucky lungs and Peter pancreas. This can cause Bronwyn bacterial lung infections leading to pneumonia while in Peter pancreas fat is not digested properly as there are fewer enzymes. This results in oily bowels, constipation and you can put on weight. Treatment involves daily Charlie chest physiotherapy, slapping hard to remove mucous. Enzymes are given with the food, as well as antibiotics and extra oxygen. Jenny genes therapy can be applied and the individual also needs to avoid smoky areas, and have a special diet.

Ageing is believed due to switching off the enzyme telomerase which adds codes to the end of Diana DNA so that at division it does not shorten every time. Extreme premature ageing is seen in the genetic disease Progeria. Here the mind is unaffected but Billy bone reabsorption occurs as well as disorders normally seen in elderly people such as arthritis etc. Also Stone man syndrome exists where Mollie muscles, Tommy tendons and Liddy ligaments start turning into Billy bone. This is often triggered by injuries where injuries are repaired by Billy bone so removal of excess Billy bone is also difficult.

Genetically engineered products are becoming more and more common i.e. putting useful Jenny genes into Bronwyn bacterial Diana DNA or cells. In humans it includes Jenny genes therapy where Diana DNA/new cells can start to provide the chemical or enzyme that was lacking. This is very useful for creating Diana DNA that can produce interferon, insulin, and antibodies. Many Jenny genes for disorders have now been identified for cancer and many exist for leukaemia. Screening tests are being developed. In the near future you will all be able to check your genetic code with a handheld device to give us your vulnerability to various diseases and disorders. Jenny genes therapy can also convert cells into anticancer cells. Genetics had been linked to behaviour problems or ability differences in schools which are not taken into account in class discipline or exams.

Your body attempts to save Diana DNA whatever happens so that you can pass your characteristics onto the next generation. All new cells are derived from stem cells, which are cells which have not yet specialised. Stem cells are being found in many places in the body and no longer need to be taken from foetuses or umbilical cords. They are found in blood, inside Billy bone even in the nasal cavity in Noddy nose where they replace those lost by a cold. Also as expected, many exist in developing embryos and the new-born. People are experimenting with stem cells to repair damaged tissue and even hope to help make themselves look or feel younger. These stem cells if injected into the blood, the eyes or organs are believed to search out damaged or dead cells and replace them. Even embryonic fluid containing these cells has been used to apply to people's faces to make them look younger. You can transplant many of the major organs and limbs, corneas and many other external forms of tissue. Cells can be grown around support material to make the shape of the tissue required or printed out on a 3D printer.

Genetic engineering is becoming more important. Some of the Jenny genes from the past are still in your Diana DNA such as those you used to use to synthesise your own vitamin C. Genetic modification can produce better animals and plants producing more efficiently food resources or chemicals needed for an ever-growing population.

Around twenty years of age it seems you lay down the foundations for your health in the future. Alcohol, drugs, lifestyle and diet can initiate minor problems which eventually become more visible and more serious when you become older.

14 AGEING AND FINAL DEATH-FRED RED CELL IS NO MORE

With age it is important to use it or lose it. Ageing as mentioned previously is linked to the fraying of ends of your chromosomes. Cell death is possibly linked with a chemical that only allows fifty cell divisions. Women are more than half the population and live longer than men but men still tend to have thicker Harry hair.

The oldest woman to exist was over one hundred and twenty two years the man a few years younger.

It is said you age and die the most between 0600-1200 hours. Ageing starts in the middle twenties and slows down in the forties. Ageing is progressive, all your steady state systems in your bodies change as does your body's structure, physiology and immune systems. Organs become less efficient and fewer hormones and enzymes are produced. By the age of eighty nearly every cell in your body has changed in some way. Ageing is a rundown eventually of many systems. If several systems cannot be corrected, especially your respiratory system which supplies oxygen to Brian brain which is controlling your body system, then temperature is affected. You can go into a coma and eventually die.

Ageing is also believed to be affected by metabolism, how much and what you eat. Temperature in hotter countries can make you age more quickly especially the parts exposed to the sun. With the storage of stem cells, especially if taken from your umbilical cord at birth there may be an opportunity as is happening now to recreate a worn out organ or inject these cell into an area so new cells can regenerate.

With all the disorders that gradually occur the elderly get used to pain. Eventually after constant pain it may not seem as bad. Permanent pain killers are used to repress pain. The body's own anaesthetic Endorphins kick in. A young body is able to compensate for minor problems if things in the body go wrong. This is not the case when you get older. Slinky skin suffers from geriatric decay. Blood can collect to cause bruising, bedsores or abscesses. Poor circulation leads to open wounds. In the elderly it is therefore important to maintain circulation, especially to prevent diabetes and gangrenous infections.

Once one thing goes wrong with one part of the body, this can have an effect on another organ or part of the body, especially ion balance. Eventually all the problems interact within the body at once and the body cannot cope. The patient may become critically ill and any treatment in hospital cannot redress the imbalance. Finally your temperature drops to a low level and the enzymes, hormones etc. cease to function and your body closes down.

As your Brian brain cells in visual cortex die they give off signals often seen as the tunnel of light in near death experiences. Eventually nearer death Brian brain cells detach and all functions are lost. You had energy stored in your cells as chemicals and this can build up and be released suddenly. It's the same spurt of energy that allows animals to escape at the last moment just as a predator captures them. When they are so exhausted they think all is lost this energy kicks in. Near death in animals and humans this energy is seen as a sudden movement. Here a person or animal can sit upright and in humans grasp your hand or in an animal even bite you. After death your Slinky skin shrinks so Harry hair and Nelly nails stand out more, they do not grow as some people think. At death there is a

195

sudden rise in temperature as Lucky lungs no longer can cool the blood down. The body then starts to sweat and you lose weight. This weight loss was thought and still in some cases to be the soul leaving the body.

15 CONCLUSION

All life, like the universe has rebelled against the second law of thermodynamics where it says everything is becoming more disorganised and energy lost. Life captures and stores energy and organises its use. Increased oxygen levels in the past helped animals increase in size, but nature has shown that where all animals have grown larger they have become vulnerable and can become extinct eventually like the dinosaurs.

My travels as a red blood cell have illustrated the complexity of your body, the variety and the interdependence of other systems on each other as well as the importance of cell function and their environments. It is a pity we cannot use the food we eat directly by incorporating these large units straight away into your bodies. Blood is a transport system to all cells where very small molecules are then rebuilt into new molecules that the body needs. Cells use Diana DNA instructions related to which part of the body the cell lives in to start to use these food molecules for repair and growth and especially for energy and heat. You have adapted our diet to eat and use all sorts of foods but still not cellulose or exoskeletons of crustacean. Many of you do not have to spend hours searching for food like other carnivores, although with the selection of foods available this may not be true.

They say an elephant is so big because its Stella stomach has to be big in order to get energy from grazing which it can do most of the day. You have also adapted your physiology and intelligence so as to live and survive in different extreme environments and move away from areas where you might not survive. At minus 20 ° centigrade for six minutes you die but some people are exposed to this for two minutes which helps to give them a smoother complexion. Sherpas have non-conductive layers of Slinky skin under their feet which act like shoes; did you have these in the past in the Ice Ages? You are looking at colonising other planets using your technology to cope in extreme conditions

You have come a long way from a group of single cells floating in a pond (if you believe in evolution but you can still think of yourself as a mass of cells but this time specialised rather like the instruments in an orchestra which work together to produce a symphonic human life. You still, like a single cell, have to keep your bodies stable by eating a healthy diet and regulating salt and water intakes.
A pharmacist may look at you as a bundle of chemicals, something that reacts or changes affected by age, when you take a pill or powder. But even this self-replicating, reactionary bundle of chemicals has come a very long way through evolution of life and ready to take on what the world can still throw at it.

I hope this book has, as well as providing a reference source, has helped to illustrate your complexity and indicated the challenges ahead in science and technology to stimulate and support the different systems and rhythms that keeps you functional. You still are like plants in that you use the energy of the sun to heat you so that on a sunny day your bodies require less energy. You also feel better on a sunny day.

You have come a long way or have you? Ignorance is shown still by some animals and is still exists in humans. The phrase bird brain comes into mind. Birds don't think very well as the middle part of their Brian brain is developed for movement. Some birds will ignore their young and not feed them if they do not open their Moby mouths. Some animals still have stepwise stages in their behaviour which if interrupted means they have to go back to repeat the initial behaviour i.e. start again. They may even treat their young as foreign beings being scared by their sudden movements.

Humans, are only one species unlike birds which have over nine thousand species and insects with their millions of species. You do not have any other species alive at present. This does not help with future evolution. So are you the last examples of your species. Did you kill off and even eat the rest of your species?

Compared to other animals you do not have a lot of species which are truly adapted to different environments. You have created artificial environments and become too dependent on manufactured foods. Luckily you created a health care system (like the NHS in the UK). Previously if anyone became ill in the family the cost of treatment could cripple their finances and people were reliant on freemasons and even druids. You have also developed selfishness, created your own identity and lost your full social potential.

You still unfortunately rely on fortune tellers and those who can see your future or even talk to the dead. Will you cope if everything disappears in a disaster? Will your old rituals become more important? Can you rebuild quickly if your homes are lost and be able to make mud huts?

Man is still a killer which is also seen in your apelike ancestors. The movement from moral control by religious based rules, to now looking for different answers is becoming more common. There is an increase in belief in other cults, the paranormal and other superstitions. One only has to look on at the number of publications on these in bookshops. Also an   increase in self-harm is evident a relic of the past and primitive tribes. This includes injecting saline to produce horns and depressions on one's head, open wound and even eyeball tattoos, piercings in and on every part of the body. This lack of social behaviour and caring not shown by mammals is frighteningly true and still exists illustrating how nature is hard and only the fittest will survive.

You have reached a state where you want all humans to survive from very premature babies to partially bionic humans. Eventually you could have chips in Brian brain and different parts of the body to keep you functioning normally, reducing the ageing processes and repairing damage. Many of your children are content to play computer rather than play other activities. E-mails, twitter and other forms of communication also restricts your activity. You can choose a partner, propose, choose a wedding ring and probably in the future get married online.

You must not isolate yourselves from the past or from what is left in the natural world around you. Are you a misfit in the game of life now? The first technological age was using tools and weapons, then industrial over the past years. With technology your Brian brains have received a huge variety of sensory input and choice. This has resulted from music devices, internet, and computer games etc. This has meant shorter response times and an early understanding of your world together with easier satisfaction of your needs.

 As Einstein suggested the key to life itself and how it fits in with the universe is in the nature around you. This key still remains to be discovered and perhaps one of you reading this book will find it. Einstein also said the man who regards his life as meaningless is not merely unhappy but hardly fit for life. Humans now need the chance to think at a higher level and produce new evolutionary ideas. People are resilient and can survive traumatic experiences with help from the past. Humans still have the ability to dream the impossible, seek the unknown and achieve greatness. One in sixty thousand people visiting Lourdes is miraculously healed. Where there is belief there is still hope as long as we Red cells keep travelling.

16 THE CAST
(No apologies for relating your name to a specific part if
the body it just happened that way-you never know this
may help you remember some of the technical terms e.g.
Edward epididymis)

Alphabetical list of players

Aden adenoids
Andy anus
Angela adrenal glands
Archie actin
Ashley arteries
Betty bladder
Billy bones
Bob the builder of Billy bone cells
Brian brain
Brenda bronchioles
Bronchi twins
Bronwyn bacteria
Candy cartilage
Carly carbohydrase
Cecilia's cilia
Cedric cerebellum
Charlie chest
Chelsea cerebrum
Christine capillaries
Danny demolition Billy bone cell
Dickie diaphragm
Diana DNA
Edward epididymus
Ellie eye
Elsie egg
Elvis pelvis
Ewan ears

Felicity fungus
Fiona fallopian tube
Fred red cell
Freda fat cell
Gillie gut
Gladys gland
Gordon gall bladder
Happy heart
Harry hair
Hazel hypothalamus
Katy kidneys
Larry loop of Henle
Lesley lymphatic system
Leslie large intestine
Liddy ligament
Lily liver
Lincoln larynx
Louis lipase
Lucky lungs
Mickey myelin sheath
Moby mouth
Mollie muscle
Monica myosin
Nora nails
Nellie nerve
Nicky nephron
Noddy nose
Olive ovary
Owen oesophagus
Patricia parathyroid
Patrick prostate gland
Penelope placenta
Percy penis
Peter pancreas
Pippa pituitary
Pluto platelets
Polly pepsin

MIKE PEARCE

Pridy protease
Rita ureter
Rosie rectum
Sally salivary gland
Slinky skin
Smartie small intestine
Speedo sperm
Spencer spinal cord
Spiro spleen
Stella stomach
Terrence Tonsils
Theodore thymus
Thelma thyroid
Tommy tendon
Tony testes
Vernon veins
Veronica virus
Victor villi
Winston white cells

ABOUT THE AUTHOR

Dr Mike Pearce is a biologist interested in behavior. As a scientist he has worked overseas as well as in this country. He was also a lecturer in Health and Social Care and Access to Health studies at a college in Canterbury Kent. Many of the facts presented here extend from his teaching. The use of people's names related to parts of the body often formed a memory aid for students as well as a means of captivating their amusement and interest.